Research Methodology for Social Sciences
An Elementary Book for Researchers

Research Methodology for Social Sciences

An Elementary Book for Researchers

Dr. Faizan Danish

Dr. Anil Bhat

SSPH

SATISH SERIAL PUBLISHING HOUSE

403, Express Tower, Commercial Complex
Azadpur, Delhi-110033 (India)
Phone : 011-27672852, Fax : 91-11-27672046
e-mail : info@satishserial.com, hkjain1975@yahoo.com
Website : www.satishserial.com

Published by :

SATISH SERIAL PUBLISHING HOUSE
403, Express Tower, Commercial Complex, Azadpur, Delhi-110033 (INDIA)
Phone : 011-27672852 Fax : 91-11-27672046
E-mail : info@satishserial.com, hkjain1975@yahoo.com

ISBN 978-93-88020-44-2

Composed, Designed & Printed in India

Preface

This book is an introduction to the basic ideas and techniques of research, statistics and is intending to prepare the students of economics, social sciences, and business management etc. for better understanding of the literature. It is now generally accepted that a study of social sciences etc. is incomplete without the knowledge of statistics. In view of increasing complexity and variety of problems, a student/scholar without a fair grounding in these tools may not be able to cope and hence, remain unfamiliar with many aspects of theories. Due to unbounded fear of so called difficult statistics tend to shy away from the subject and often, it may be difficult to persuade them to take special courses in statistics. Statistics is best learnt if the student handles it systematically. He is bound to find it rewarding and exciting while following the logical arguments and steps involved. This book gives a simple and clear presentation of logical and meaning of statistics which will be useful to beginners. There is fairly self-contained development of topics and the student would find here a starting point which will help gain self confidence and familiarity with the subject. Quite frequently these days people talk of research, both in academic institutions and outside. Several research studies are undertaken and accomplished year after year. But in most cases, very little attention is paid to an important dimension relating to research, namely, that of research methodology. The result is that much of research, particularly in social sciences, contains endless word-spinning and too many quotations. Thus, a great deal of research tends to be futile. It may be noted, in the context of planning and development that the significance of research lays in its quality and not in quantity. The need, therefore, is for those concerned with research to pay due attention to designing and adhering to the appropriate methodology throughout for improving the quality of research. The methodology may differ from problem to problem, yet the basic approach towards research remains the same. Keeping all this in view, the present book has been written with two clear objectives, viz., (i) to enable researchers, irrespective of their discipline, in developing the most appropriate methodology for their research studies; and (ii) to make them familiar with the art of using different research methods and techniques. It is hoped that the humble effort made in the form of this book will assist in the accomplishment of exploratory as well as result-oriented research studies. Regarding the organization, the book consists of fourteen chapters, well-arranged in a coherent manner.

Dr. Faizan Danish
Dr. Anil Bhat

About the Authors

Dr. Faizan Danish is working as a Lecturer in Statistics at Department of Mathematics and Statistics, Government Degree College, Shopian, J&K India. He has done Doctor of Philosophy in Statistics from Division of Statistics and Computer Science, Faculty of Basic Sciences, Sher-e-Kashmir University of Agricultural Sciences and Technology Jammu, J&K, India and Post Graduation from Department of Statistics, University of Kashmir, Srinagar, J&K, India. His research interests are Sampling theory, Mathematical Programming and Applied Statistics. He is recipient of the IMRF Young Scientist Award for his incredible contribution towards the research and has published research articles in reputed journals of Statistics and Mathematics. Not only had this, but also he has published a book and several chapters in different areas of Statistics. He has presented a number of papers in several national as well as international conferences.

Dr. Anil Bhat, Assistant Professor, Agricultural Economics, Division of Agricultural Economics & ABM, SKUAST-Jammu possesses a meritorious academic record. He has done M.Sc. (Agril. Economics) from the Chaudhary Charan Singh University, Meerut and Ph.D. (Agril. Economics) from the Sher-e-Kashmir University of Agricultural Sciences and Technology of Jammu. He also holds the Post Graduate Diploma in Rural Development from Indira Gandhi National Open University, New Delhi and Post Graduate Diploma in Agricultural Marketing from Pondicherry University, Kalapet, Puducherry. He is the recipient of certificate of "SPS measures under WTO" from WTO, Geneva, Switzerland. He has 50 publications including 12 books chapters and seven books to his credit. He has guided 5 research students (M. Sc. and MBA Agri-Business Management) as Major Advisor. He is the member of four professional societies and is also reviewer and editor of professional journals. Besides, he is also the recipient of Young Associate Scientist Award and Young Scientist Award in the field of Agricultural Economics and has also received number of paper presentation awards. His field of expertise is production, marketing and agri-business management.

Contents

Preface .... *v*

1. Measures of Location 1
   Dr. Tanveer Ahmad Tarray

2. Measures of Dispersion 21
   Dr. Tanveer Ahmad Tarray

3. Correlation 37
   Dr. Faizan Danish, Yasmeena Ismail and Mir Subzar

4. Regression Analysis 53
   Firdoos Yousuf

5. Sampling Theory 67
   Dr. Faizan Danish

6. Economics of Production of Capsicum (Capsicum Annum) in Solan District of Himachal Pradesh 81
   Subhash Sharma, Sangya Kumari, Amit Guleria, RS Prasher, MK Vaidya and Ravinder Sharma

7. Economics of Production of Mango: A Case Study of Kangra District of Himachal Pradesh 95
   Subhash Sharma, Parul Sharma, Amit Guleria, RS Prasher, MK Vaidya and Ravinder Sharma

8. Research Methodology for Social Sciences 115
   Manpreet Kaur and Dr. Anil Bhat

9. Investigating the Role of Mediation Analysis in Social Sciences Research 127
   Shelleka Gupta

Chapter 1

Measures of Location

Dr. Tanveer Ahmad Tarray

Department of Mathematical Sciences, Islamic University of Science and Technology, Awantipoora-192122 (J&K), India, E-mail: tanveerstat@gmail.com

1. Frequency Distributions, Diagrams and Graphs

1.1 Introduction

Statistics, as it is today, is very closely related and very much indebted to Mathematics. Wide applications of advanced Mathematics have resulted into recent advancements in statistical analysis. Thus, statistics may be regarded as that branch of Mathematics which systematically analyses large number of related numerical facts. It is that branch of Mathematics which specializes in data. Therefore, arises the need for a new branch of Mathematics called mathematical Statistics.

Statistical data, discrete or continuous, available in an arbitrary manner from the field of enquiry on a single characteristic of a large number of individuals are called raw data. These data will not provide any useful information of interest unless we classify and arrange them in proper tabular form. Let us consider the marks obtained by 60 students to a class in statistics:

Table 1.1 Marks in Statistics of 60 students

55	15	18	25	56	39	26	18	32	15	25	25	22	25	46
25	02	36	35	35	68	35	32	38	56	32	22	46	48	10
42	36	24	64	39	35	64	42	45	40	56	48	18	78	42
47	54	50	20	63	45	35	26	54	58	35	68	46	75	54

Data presented in the above table are called 'raw data'. This type of representation of data will not furnish any useful information and is rather confusing to mind to a statistician. If the marks of the students given in the above table are arranged in ascending or descending order of magnitude commonly known as 'array' of the data but this does not reduce the bulk of the data.

A bar (|) called 'tally' mark is placed against the number, when it occurs. Having occurred four times, the fifth occurrence is represented by putting a cross tally (\) on the first four lines. This technique facilities the counting of the tally marks at the end. Marks are called the variable (x) and the number of students against the marks is known as the 'frequency (f)' of the variable. The word 'frequency' is derived from 'how frequently' a variable occurs. For example, in the above case the frequency of 25 is 5 as there are five students getting 25 marks. Such a frequency distribution is known as 'simple' or 'ungrouped' frequency distribution. For the above data, the simple frequency distribution is given below. A tabular arrangement of the data techniques showing the frequency with which each successive value of the variable occurs is called a 'frequency distribution'. The data, which are not in a tabular form like the ungrouped data or the array. We shall call a 'simple distribution'.

Table 1.2 Simple frequency distribution of marks of 60 students:

Marks	Marks	No. of Students
2	|	1
10	|	1
15	|	2
18	|	2
20	|	1
–	–	–
–	–	–
Total	–	**60**

1.2 Grouped Frequency Distribution

The representation of data as a simple frequency distribution is better than array, but it does not reduce the bulk of the data much and such the frequency table remains too lengthy. The eye and mind cannot quickly look over all the sets of observations. The next step is, therefore, to condense the data by putting them into smaller groups or classes called 'class interval' mentioned earlier. The number of values belonging to each class interval is called the class frequency. Such a table showing the distribution of the frequencies in the different classes is called 'frequency table' and the manner in which the class frequencies are distributed over the class interval is called the 'grouped frequency distribution' of the variable. For example, in the above case the data may be expressed as shown in Table 1.3. This type of presentation of data is very much popular in practice.

Table 1.3 Marks obtained by 60 students of a class in Statistics

Marks (x)	Number of Students (f)
01-10	2
11-20	6
21-30	10
31-40	15
41-50	12
51-60	8
61-71	5
71-80	2
Total	60

1.3 Continuous Frequency Distribution

If we deal with a continuous variable, it is not possible to arrange the data in the class intervals of above type. Let us consider the distribution of age in years, if the class intervals are 15-19, 20-24, etc., then the persons with ages between 19 and 20 years are not taken into consideration. In such a case, we form the class intervals as shown below:

Age in years,

Below 5,

5 or more but less than 10,

10 or more but less than 15,

15 or more but less than 20,

20 or more but less than 25,

and so on.

Here, all the persons with any fraction of age are included in one group or other. For practical points of view, we re - write the above classes as:

0-5, 5-10, 10-15, 15-20, 20-25 and so on.

This form of frequency distribution is known as 'continuous frequency distribution'.

In Table 3.3, frequency distribution is discrete.

Table 1.4 Marks obtained by 60 students of a class in Statistics

Marks (x)	Number of Students (f)
0.5-10.5	2
10.5-20.5	6
20.5-30.5	10
30.5-40.5	15
40.5-50.5	12
50.5-60.5	8
60.5-70.5	5
70.5-80.5	2
Total	60

1.4 Relative Frequency Distribution

If $f_1, f_2, f_3, \ldots, f_k$ are frequencies, then $\frac{f_1}{N}, \frac{f_2}{N}, \frac{f_3}{N}, \ldots, \frac{f_k}{N}$ are respectively the relative frequencies of the different classes. It is obvious that the sum of all the relative frequencies is equal to one. Such distribution is known as 'relative frequency distribution'. Let us consider an example of a relative frequency distribution.

Table 1.5 The frequency distribution of 100 persons with relative frequency

Age in years (x)	Number of students (f)	Relative frequency (f/N)
0-9	2	0.05
10-19	6	0.15
20-29	10	0.20
30-39	15	0.25
40-49	12	0.15
50-59	8	0.12
60-69	5	0.08
Total	**100**	**1.00**

Relative frequency distributions are used most commonly in connection with the topic of probability.

1.5 Comulative Frequency Distribution

A frequency distribution simply tells us, how frequency a particular value of the variable is occurring. However, if we want to know the total number of members getting a value 'less than' or 'more than' a particular value of the variable, this frequency table fails to prove the information as such .The information can merely be obtained very conveniently from cumulative frequency distribution which is a modification of the given frequency distribution and is obtained on successively adding the frequencies of the values of the variable according to a certain law. The frequencies so obtained are called the ' cumulative frequency' and is denoted by c.f., if the frequencies are cumulated from top of the table giving frequencies 'less than' type frequency c.f., we shall explain the construction of such distribution by means of a numerical example.

Table 1.6 Distribution showing the c.f. less than and c.f. more than types from simple frequency

Marks (x)	Number of Students (f)	C.F. Less than	C.F. More than
30-35	5	5	5+65=70
35-40	10	10+5=15	5+60=65
40-45	15	15+15=30	30+30=60
45-50	30	30+30=60	15+15=30
50-55	5	5+60=65	10+5=15
55-60	5	5+65=70	5
Total	70		

A table such as above presenting cumulative frequencies, is called a 'cumulative frequency distribution'. If a question is asked that how many students are getting the marks less than 35, less than 40, less than 45 and less than 50 marks, etc., then the answer is 5, 15, 30 and 60 students respectively. For more than 30, more than 35, more than 40 and more than 45 marks, etc., the answer is 70, 65, 55 and 40 students respectively. This clearly indicates that cumulative frequency of less than type always corresponds to upper limit of the class intervals and cumulative frequency of more than type always corresponds to lower limit of the class interval. If the relative frequencies are cumulated from the top of the table, it is called ' cumulative relative frequency distribution'.

1.6 Frequency Density

If we divide the frequency in a class of distribution by the width of the class interval, we obtain what is called 'frequency density' i.e.

$$\text{Frequency Density} = \frac{\text{Frequency in the class interval}}{\text{Width of the class interval}}$$

In other words, frequency density is frequency per unit of interval. If the classes of frequency distribution are not of equal width, we can not compare the frequency of the different classes. In this case , frequency density of the classes is used for comparing the concentration of frequencies in the classes.

1.7 Relative Frequency Density

If we divide the 'frequency density' by the population size N, we obtain what is called 'relative frequency density' i.e.

$$\text{Relative Frequency Density} = \frac{\text{Frequency Density}}{\text{Population Size (N)}}$$

1.8 Variable and Variates

A quantity which can vary from one individual to another is called a variable or variate, e.g., height, weight, ages, wages of persons, rainfall records of cities, etc. The variate is a little bit different from a variable. For example, the number of inches or centimeters that represents the height of an individual is a variate whereas the height is its variable. Quantities which can take any numerical value within a certain range are called continuous variable e.g., as the child grows, his/her height takes all possible values from 50 cm to 100 cm. Quantities which are incapable of taking all possible values are called discrete or discontinuous variable e.g., the number of children a man can have are +ve integers 1,2,3 etc. (no value between any two consecutive integers).

Table 1.8 Make a frequency table having grades of wages with class interval of Rs. 2 each from the following data of daily wages received by 30 workers in a certain factory. Daily wages in rupees are:

14,16,16,14,22,13,15,24,12,23,14,20,17,21,22,18,18,19,20,17, 16,15,11,12,21,20,17,18,19,23.

Solution: Minimum entry of raw data = 11, width of a class interval = 2 (given)

Wages in Rs.	Number of Workers	Frequency
11-13	III	3
13-15	IIII	4
15-17	卌	5
17-19	卌 I	6
19-21	卌	5
21-23	III	4
23-25	III	3
Total	-	30

In the study of a population with respect to one in which we are interested, we may get a large number of observations. It is impossible to get any idea about the characteristic when we look at all the observations. So, it is suggested to have one number for one group. That number have to be a good representative one for all the values to give a clear picture of that characteristic. Such a number can be a central value for all these values. Such a central value is known as measure of central tendency or an average or a measure of locations. There are five averages named as Mean, Median, Mode, Harmonic Mean and Geometric Mean, in which mean, median and mode are called simple averages and the other two averages geometric mean and harmonic mean are called special averages.

A good or ideal average has certain characteristics. The following properties should possess for an ideal average:

1. It should be rigidly defined.
2. It should be simple to get and compute.
3. It should be based on all items in the data.
4. Its definition shall be in the form of a mathematical formula.
5. It should be capable of further algebraic treatment.
6. It should have sampling stability.
7. It should be capable of being used in further statistical computations or processing.

1.9 Arithmetic mean or mean

Arithmetic mean or simply the mean of a variable is defined as the sum of the observations divided by the number of observations. If the variable x assumes n values $x_1, x_2, \ldots, x_n$, then the mean, $\bar{x}$ is given by

$$\bar{x} = \frac{x_1 + x_2 + \ldots\ldots + x_n}{n} = \frac{1}{n}\sum_{i-1}^{n} x_i$$

This formula is for the ungrouped or raw data.

Example 1: Calculate the mean for 2, 4, 6, 8, 10

Solution: We have

$$\bar{x} = \frac{1}{n}\sum_{i-1}^{n} x_i$$

$$\therefore \bar{x} = \frac{1}{5}(2+4+6+8+10) = \frac{30}{5} = 6$$

Merits and demerits of Arithmetic mean

Merits

1. It is unbendingly characterized.
2. It is easy to comprehend and simple to figure.
3. In the event that the quantity of things is adequately substantial, it is more precise and more dependable.
4. It is a figured value and did not depend on its situation in the series.
5. It is conceivable to ascertain regardless of whether a portion of the points of interest of the information are inadequate.
6. Everything being equal, it is influenced slightest by changes of inspecting.
7. It gives a decent premise to examination.

Demerits

1. It can neither be acquired by examination nor situated through a recurrence chart.

2. It can't be in the investigation of subjective wonders not fit for numerical estimation i.e. Insight, magnificence, genuineness and so on.
3. It can disregard any single thing just at the danger of losing its exactness.
4. It is influenced particularly by outrageous qualities.
5. It can't be figured for open-end classes.
6. It might prompt fraudulent ends, if the subtle elements of the information from which it is figured are not given.

1.10 Median

In the words of L.R. Conner : "The median is that value of the variable which divides the data in two equal parts, one part comprising all the values greater and the other, all values less than median." Thus, as against arithmetic mean which is based on all the items of the distribution, the median is only positional average, i.e. the value depends on the position occupied by a value in the frequency distribution.

Computation of Median

Ungrouped data: If the number of observation is odd, then the median is the middle value after the observations have been arranged in ascending or descending order of magnitude. In case of even number of observations, median is obtained as the arithmetic mean of two middle observations after they are arranged in ascending or descending order of magnitude.

Problem: The marks obtained by 12 students out of 50 are: 25, 20, 23, 32, 40, 27, 30, 25, 20, 10, 15, and 41

Solution: The values obtained by 12 students arranged in ascending order as:

10, 15, 20, 20, 23, 25, 25, 27, 30, 32, 40, 41

Here the number of items 'N' = 12, which is even.

The two middle items are 6th and 7th items

i.e., $\frac{25+25}{2} = 25$ is the median value.

(Case 11) Frequency (Discrete) Distribution: In case of frequency distribution where the variables take the value X1, X2, ...,Xn with respective frequencies f1, f2, ..., fn with $N = \sum_{i=1}^{n} f_i$, median is the size of the $\frac{(N+1)}{2}th$ item or observation. In this case, the use of cumulative frequency (c.f.) distribution facilitates the calculations. The steps involved are:

i) Prepare the less than cumulative frequency (c.f.) distribution.

ii) Find N/2.

iii) Find the c.f. just greater than N/2.

iv) The corresponding value gives the median.

Problem: From the following data find the value of median:

Income (Rs.)	1000	1500	800	2000	2100	1700
No. of Persons	24	26	14	10	5	28

Solution:

Income arranged in ascending order	**No. of Persons(f)**	**c.f.**
800	14	14
1000	24	38
1500	26	64
1700	28	92
2000	10	102
2100	5	107

Median = Size of (N/2)th item = 107/2 = 53.5. The c.f just greater than 53.5 is 64. The corresponding value to this = 1500. Hence, Median = Rs. 1500.

(Case 111) Continuous Frequency Distribution: Steps involved for its computation are:

i) Prepare less than cumulative frequency (c.f.) distribution.

ii) Find N/2.

iii) Locate c.f. just greater than N/2.

iv) The corresponding class contains the median value and is called the median class.

v) The value of median is now obtained by using the interpolation formula:

$$mediam = L + \frac{h}{f}\left(\frac{N}{2} - C\right)$$

Where L is the lower limit or boundary of the median class;

f is the frequency of the median class;

h is the magnitude or width of class interval;

is the total frequency, and

C is the cumulative frequency of the class preceding the median class.

Problem: The annual profits (in Rs. lacs) shown by 60 firms are given below:

Profits:	15-20	20-25	25-30	30-35	35-40	40-45	45-50	50-55	55-60	60-65
No. of Firms:	4	5	11	6	5	8	9	6	4	2

Calculate the median.

Solution:

Profits	No. of Firms(f)	Cumulative frequency (c.f.)
15-20	4	4
20-25	5	9
25-30	11	20
30-35	6	26
35-40	5	31
40-45	8	39
45-50	9	48
50-55	6	54
55-60	4	58
60-65	2	60

Median item = N/2 =60/2= 30

The cumulative frequency just greater than 30 is 31 and is corresponding class 35-40 is the median class.

$$mediam = L + \frac{h}{f}\left(\frac{N}{2} - C\right)$$

$$\therefore mediam = 35 + \frac{5}{5}\left(\frac{60}{2} - 26\right)$$

$$= 35 + 4 = 39 marks$$

Merits and Limitations of Median

The median is superior to arithmetic mean in certain aspects. For example, it is especially useful in case of open-ended distribution and also it is not influenced by the presence of extreme values. In fact when extreme values are present in a series, the median is more satisfactory measure of central tendency than the mean.

However, since median is positional average, its value is not determined by each and every observation. Also, median is not capable of algebraic treatment. For example, median cannot be used for determining the combined median of two or more groups. Furthermore, the median tends to be rather unstable value, if the number of observations is small.

1.11 Mode

Mode is the value which occurs most frequently in the set of observations. It is the point of maximum frequency or the point of greatest density. In other words, the mode or modal value of the distribution is that value of the variate for which frequency is maximum.

Computation of Mode

a) In case of discrete frequency distribution, mode is the value of the variable corresponding to the maximum frequency.

   But in any one (or more) of the following cases:

   i) If the maximum frequency is repeated

   ii) If the maximum frequency occurs in the very beginning or at the end of the distribution

   iii) If there are irregularities in the distribution, the value of mode is determined by the method of grouping.

b) In case of continuous frequency distribution, mode is given by the formula :

Where, l is lower limit, h the width and f_m the frequency of the modal class; f_1 and f_2 are the frequencies the classes preceding and succeeding the modal class respectively.

Note:

1. For symmetric distribution mean, mode and median coincide.
2. When mode is ill defined i.e., where the method of grouping also fails, its value is ascertained by the formula

Mode=3 Median – 2 Mean

Example : If seven men are receiving daily wages of Rs. 150, 160, 170, 170, 170, 180 and 200. Find the modal wages.

Solution : Since 170 occurs thrice and no other item occurs three times or more than three times, hence modal wages is Rs. 170.

Example: Determine the mode from the following frequency distribution:

X	1	2	3	4	5	6	7	8
F	4	9	16	25	22	16	8	3

Solution : Here maximum frequency is 25 and the corresponding value of X is 4. Hence, mode is 4.

Example : Calculate the mode from the following frequency distribution:

Size (x)	4	5	6	7	8	9	10	11	12	13
Frequency (f)	2	5	8	9	12	14	14	15	11	13

Solution: Method of grouping

<table>
<tr><th rowspan="2">Size (x)</th><th colspan="6">Frequency</th></tr>
<tr><th>I</th><th>II</th><th>III</th><th>IV</th><th>V</th><th>VI</th></tr>
<tr><td>4</td><td>2</td><td rowspan="2">7</td><td rowspan="1"></td><td rowspan="3">15</td><td rowspan="1"></td><td rowspan="2"></td></tr>
<tr><td>5</td><td>5</td><td rowspan="2">13</td><td rowspan="3">22</td></tr>
<tr><td>6</td><td>8</td><td rowspan="2">17</td><td rowspan="3">29</td></tr>
<tr><td>7</td><td>9</td><td rowspan="2">21</td><td rowspan="3">35</td></tr>
<tr><td>8</td><td>12</td><td rowspan="2">26</td><td rowspan="3">40</td></tr>
<tr><td>9</td><td>14</td><td rowspan="2">28</td><td rowspan="3">43</td></tr>
<tr><td>10</td><td>14</td><td rowspan="2">29</td><td rowspan="3">40</td></tr>
<tr><td>11</td><td>15</td><td rowspan="2">26</td><td rowspan="3">39</td></tr>
<tr><td>12</td><td>11</td><td rowspan="2">24</td><td rowspan="2"></td></tr>
<tr><td>13</td><td>13</td><td></td></tr>
</table>

- In column I, original frequencies are written.
- In column II, frequencies of column I are combined two by two.
- In column III, leave the first frequency of column I and combine the others two by two.
- In column IV, frequencies of column I are combined three by three.
- In column V, leave the first frequency of column I and combine the others three by three.
- In column VI, leave the first two frequencies in column I and combine the others three by three.

Now, we frame another table in which against every maximum item in column I to VI, we write down the corresponding size or sizes. The size (x) which occurs maximum number of times is the mode.

Columns	Size of items having maximum frequency
I	11
II	10,11
III	9,10
IV	10,11,12
V	8,9,10
VI	9,10,11

Since the item 10 occurs maximum number of times i.e., 5 times, hence the mode is 10.

Merits and Demerits of Mode

Merits

1. It is promptly understandable and simple to compute.
2. It tends to be situated now and again by investigation.
3. In the event that isn't affected by extraordinary qualities.
4. It depends on every one of the estimations of the variate.

Demerits

1. It is sick - characterized . An unmistakably characterized mode does not generally exist.
2. It did not depend on every one of the perceptions.
3. It isn't prepared to do facilitate numerical medicines.

1.12 Geometric Mean

a) Geometric mean (G.M.) of n individual observations $x_1, x_2, ...x_n (x_i \neq 0)$ is the nth root of their product.

Thus, $G = (x_1, x_2, ...x_n)^{1/n}$

Taking logarithms of both sides

$$LogG = \frac{1}{n}(\log x_1 + \log x_2 + ... + \log x_n)$$

$$= \frac{1}{n}\sum_{i-1}^{n} \log x_i$$

$$G = anti\log\left[\frac{1}{n}\sum_{i=1}^{n} \log x_i\right]$$

b) If $x_1, x_2, ...x_n (x_i \neq 0)$ occur $f_1, f_2, ...f_n$ times respectively and $N = \sum_{i-1}^{n} f_i$ then G.M. is given by

Thus, $G = (x_1^{f_1} x_2^{f_2}, ...x_n^{f_n})^{1/n}$

Taking logarithms of both sides

$$LogG = \frac{1}{n}(f_1 \log x_1 + f_2 \log x_2 + ... + f_n \log x_n)$$

$$\frac{1}{n}\sum_{i-1}^{n} f_i \log x_i$$

$$G = anti\log\left[\frac{1}{n}\sum_{i=1}^{n} f_i \log x_i\right]$$

c) In case of continuous frequency distribution, x is taken to be the value corresponding to the mid-points of the class – intervals.

Example : Compute the geometric mean from the following data:

10, 110, 120, 550, 80, 0, 52, 37

Solution:

Size of items (x)	Log x
10	1.0000
110	2.0414
120	2.0792
50	1.990
80	1.9031
0	1.7782
52	1.7160
37	1.582
No. of items (n)=8	**13.7851**

$$\therefore LogG = \frac{1}{n}\sum_{i=1}^{n} \log x_i = \frac{1}{8}(13.7851) = 1.723$$

$$\therefore G = anti\log(1.723) = 52.84$$

Example: Compute the geometric mean from the following data:

Marks	No. of Students
0-10	10
10-20	5
20-30	8
30-40	7
40-50	20

Solution:

Marks	Mid-values (x)	No. of Students (f)	log x	f log x
0-10	5	10	0.6990	6.9900
10-20	15	5	1.1761	5.8805
20-30	25	8	1.3979	11.1832
30-40	35	7	1.5441	10.8087
40-50	45	20	1.5532	33.040
		N=50		67.924

$$\therefore LogG = \frac{1}{N}\sum_{i=1}^{n} f \log x_i = \frac{1}{50}(7.9264) = 1.3585$$

$$\therefore G = anti\log(1.3585) = 22.83$$

Merits and Demerits of Geometric Mean

Merits

1. It is inflexibly characterized.
2. It is very little influenced by variance of examining.
3. It depends on every one of the perceptions.

Demerits

1. For a non-scientific understudy, it is difficult to comprehend and compute .
2. On the off chance that any of the perceptions is zero, G.M. is zero and if an odd number of perceptions are negative, it can't be computed.

1.13 Harmonic Mean

Harmonic mean of a number of observations is the reciprocal of the arithmetic mean of the reciprocals of the given values. Thus, the harmonic mean H of n observations $x_1, x_2, ...x_n (x_i \neq 0)$

$$H = \frac{1}{\frac{1}{n}\sum_{i=1}^{n}\frac{1}{x_i}} = \frac{n}{\frac{1}{x_1} + \frac{1}{x_2} + ... + \frac{1}{x_n}}$$

If $x_1, x_2, \ldots\ldots x_n (x_i \neq 0)$ occur $f_1, f_2, \ldots\ldots f_n$ times respectively, then harmonic mean is given by:

$$H = \frac{1}{\frac{1}{N}\sum_{i=1}^{n}\frac{f_i}{x_i}}$$

$$H = \frac{N}{\frac{f_1}{x_1}+\frac{f_2}{x_2}+\ldots..+\frac{f_n}{x_n}}, N = \sum_{i=1}^{n} f_i$$

In case of continuous frequency distribution, x is taken to be the value corresponding to the mid-points of the class – intervals.

Example: Compute the harmonic mean of the following: 5, 10, 20

Solution:

$$H.M = \frac{3}{\frac{1}{x_1}+\frac{1}{x_2}+\ldots.+\frac{1}{x_n}} = \frac{3}{\frac{1}{5}+\frac{1}{10}+\ldots.+\frac{1}{20}} = 8.57$$

Example: Find out the harmonic mean of the following data:

Marks (out of 150)	No. of Students
10	2
20	3
40	6
60	5
120	4
	N=20

Solution:

x	f	1/x	f/x
10	2	.100	.200
20	3	.050	.150
40	6	.025	.150
60	5	.017	.085
120	4	.008	.032
	N=20		**.617**

$$\therefore H = \frac{N}{\sum_{i=1}^{n} \frac{f_i}{x_i}} = \frac{20}{.617} = 32.4$$

Example: Compute the harmonic mean from the following data:

Marks	No. of Students
0-10	4
10-20	6
20-30	10
30-40	7
40-50	3

Solution:

Marks	Mid-values (x)	No. of Students (f)	1/x	f/x
0-10	5	4	.200	.800
10-20	15	6	.067	.402
20-30	25	10	.040	.400
30-40	35	7	.029	.203
40-50	45	3	.022	.066
		N=30		**1.871**

$$\therefore H = \frac{N}{\sum_{i=1}^{n} \frac{f_i}{x_i}} = \frac{30}{1.871} = 16.03$$

Merits and demerits of Harmonic Mean

Merits

1. It is unbendingly characterized.
2. It is appropriate for advance numerical treatment.
3. It depends on every one of the perceptions.

Demerits

1. It is hard to figure.
2. It is hard to get it.

References

1. Anderson B.J.A. (1989). Interpreting Data: A First Course in Statistics, Chapman and Hall, London.
2. Akaike H. (1974). A New Look at the Statistical Model Identification, IEEE transactions on Automatic Control, 19: 716-723.
3. Chatfield C. (1996). The Analysis of Time Series,London:Chapman and Hall.
4. Clarke M.G. and Cooke D. (2004).A Basic Course in Statistics, 5th Edition, Arnold, London.
5. Converse J. M. and Presser S. (1986). Survey Questions: Handcrafting the Standardized Questionnaire, London: Sage.
6. Coles S. (2001). An Introduction to Statistical Modelling of Extreme Values, Springer, London.
7. Collett D. (2003). Modelling Survival Data in Medical Research, 2nd Edition, Chapman and Hall/CRC, Boca Raton, FL.
8. Crowder J.M. and Hand J.D. (1990).Analysis of Repeated Measures. Chapman and Hall, London.
9. Daly F., Hand J.D., Jones M.C., Lunn D. A. and McConway J. K. (1995). Elements of Statistics,The Open University and Addison-Wesley, Wokingham.
10. Diamond and Jefferies (2001). Beginning Statistics: An Introduction for Social Scientists, Sage Publications, London.
11. Dielman T. E. (1996). Applied Regression Analysis for Business and Economics, 2nd edition, Belmont, California: Duxbury Press.
12. Everitt S. B. (1999).Chance Rules: An Informal Guide to Probability, Risk and Statistics, Springer-Verlag, New York.
13. Feller W. (1971). An Introduction to Probability Theory and its Applications, VolumeII., Wiley, New York.
14. Fisher R.A. and Yates F. (1938). Statistical Tables for Biological, Agricultural, and Medical Research, Oliver and Boyd, Edinburgh.

15 Goon A.M., Gupta M.K. and Dasgupta B. (2001). Fundamentals of Mathematical Statistics (Vol.2), Word Press

16 Hand J. D. (1997).Construction and Assessment of Classification Rules, John Wiley & Sons Ltd., Chichester.

17 Danish, F. (2017). Fundamentals of Statistics: A Brief Insight, Elite Publishers, Delhi.

❑❑❑

Chapter 2

Measures of Dispersion

Dr. Tanveer Ahmad Tarray

Department of Mathematical Science, Islamic University of Science and Technology, Awantipoora-192122, (J&K), India, E-mail: tanveerstat@gmail.com

2.1 Measures of Dispersion

The measures of central tendency serve to find the center of the distribution, yet they don't uncover how the things are spread out on either side of the middle. This normal for frequency distribution is generally alluded to as scattering. In an arrangement, every one of the things are not equivalent. There is distinction or variety among the qualities. The level of variation is assessed by different proportions of scattering. Small dispersion demonstrates high consistency of the things, while huge dispersion shows less uniformity. For instance, think about the accompanying characteristics of two students.

Student I	Student II
68	85
75	90
65	80
67	25
70	65

Both have got a total of 345 and an average of 69 each. The fact is that the second student has failed in one paper. When the averages alone are considered, the two students are equal. But first student has less variation than second student. Less variation is a desirable characteristic.

Characteristics of a Good Measure of Dispersion

A perfect proportion of variation is relied upon to have the accompanying properties:

1. It ought to be unbendingly characterized.

2. It ought to be based on every one of the things.
3. It ought not be unduly influenced by extreme values.
4. It ought to lend itself for mathematical manipulation.
5. It ought to be easy to comprehend and simple to calculate.

Absolute and Relative Measures

There are two sorts of measures of dispersion, in particular absolute measure of dispersion and relative measure of dispersion. Absolute measure of dispersion shows the measure of variation in an arrangement of values in terms of units of observation. For instance, when rainfalls on various days are accessible in mm, any absolute measure of dispersion gives the variation in rainfall in mm. Then, again relative measures of dispersion are free from the units of measurements of the observation. They are unadulterated numbers. They are utilized to analyze the variation in at least two sets, which are having diverse units of measurements of variation. The different absolute and relative measures of dispersion are recorded beneath.

Absolute measure	Relative measure
Range	Co-efficient of Range
Quartile deviation	Co-efficient of Quartile deviation
Mean deviation	Co-efficient of Mean deviation
Standard deviation	Co-efficient of Variation

2.2 Range and Co-efficient of Range

Range : The range is the simplest measure of dispersion. It is the rough measure of dispersion. Its measure depends upon the extreme items and not on all the items. It is defined as the difference between the largest and smallest values of the variable.

In symbols, Range = L – S.

Where L = Largest value.

S = Smallest value.

In individual observations and discrete series, L and S are easily identified. In continuous series, the following two methods are followed:

Method 1

L = Upper boundary of the highest class.

S = Lower boundary of the lowest class.

Method 2

L = Mid-value of the highest class.

S = Mid-value of the lowest class.

Co-efficient of Range

$$\text{Co-efficient of Range} = \frac{L - S}{L + S}$$

Example 1 : Find the value of range and its co-efficient for the following data 7, 9, 6, 8, 11, 10, 4

Solution: Here L=11 and S = 4.

We have

Range = L – S = 11- 4 = 7

Also, $$\text{Co-efficient of Range} = \frac{L - S}{L + S}$$

$$\therefore \quad \text{Co-efficient of Range} = \frac{11 - 4}{11 + 4} = \frac{7}{15} = 0.4667$$

Example 2 : Calculate range and its co-efficient from the following distribution:

Size:	60-63	63-66	66-69	69-72	72-75
Number:	5	18	42	27	8

Solution:

L = Upper boundary of the highest class = 75

S = Lower boundary of the lowest class = 60

We have

Range = L – S = 75 – 60 = 15

Also, $\text{Co - efficient of Range} = \frac{\text{L - S}}{\text{L + S}}$

$$\therefore Co - efficient\ of\ Range = \frac{75-60}{75+60} = \frac{15}{135} = 0.1111$$

Merits and Demerits of Range

Merits

1. It is easy to get it.
2. It is anything but difficult to figure.
3. In certain sorts of issues like quality control, climate conjectures, share value examination, and so forth, range is most broadly utilized.

Demerits

1. It is especially influenced by the extreme values.
2. It depends on just two extreme observations.
3. It can't be ascertained from open-end class intervals.
4. It isn't reasonable for mathematical treatment.
5. It is once in a while utilized measure.

2.3 Quartile Deviation

The difference between the upper and lower quartiles i.e., Q_3-Q_1 is known as the inter quartile range and half of it i.e., $\frac{1}{2}(Q_3 - Q_1)$ is called the semi-inter quartile range or the quartile deviation (Q.D).

$$Quartile\ deviation = \frac{1}{2}(Q_3 - Q_1)$$

It is better measure of dispersion than range. By eliminating the lowest 25% and highest 25% of items in a series, we are left with the central 50%, which are ordinarily free of extreme values.

$$Co - efficient\ of\ Quartile\ deviation = \frac{Q_3 - Q_1}{Q_3 + Q}$$

Example : Find the Quartile Deviation and Coefficient of Quartile Deviation for the following data: 391, 384, 591, 407, 672, 522, 777, 733, 1490, 2488

Solution: Arrange the given values in ascending order. 384, 391, 407, 522, 591, 672, 733, 777, 1490, 2488.

We have $Q_1 = \left(\frac{N+1}{4}\right)^{th}$ *item*

$$\therefore Q_1 = \left(\frac{10+1}{4}\right) = 2.75^{th}\, item$$

Q1 = 2nd value + 0.75 (3rd value - 2nd value)

= 391 + 0.75 (407 - 391)

= 391 + 12

= 403

We have $Q_3 = 3\left(\frac{N+1}{4}\right)^{th}$ *item*

$$\therefore Q_3 = 3\left(\frac{10+1}{4}\right) = 8.25^{th}\, item$$

Q3 = 8th value + 0.25 (9th value - 8th value)

= 777 + 0.25 (1490 - 777)

= 777 + 178.25

= 955.25

We know that

$$\text{Quartile deviation} = \frac{1}{2}(Q_3 - Q_1)$$

$$\therefore \text{Quartile deviation} = \frac{1}{2}(955.25 - 403) = \frac{552.25}{2} = 276.12$$

Also, $\text{Co-efficient of Quartile deviation} = \frac{Q_3 - Q_1}{Q_3 + Q_1}$

$$\therefore \text{Co-efficient of Quartile deviation} = \frac{955.25 - 403}{955.25 + 403} = \frac{552.25}{1358.25} = 0.41$$

Example : Calculated the Quartile Deviation and Coefficient of Quartile Deviation from the following data:

Age in years:	20	30	40	50	60	70	80
No. of members:	3	61	132	153	140	51	3

Solution:

Age in years	No. of members	c.f
20	3	3
30	61	64
40	132	196
50	153	349
60	140	489
70	51	540
80	3	543
Total	**N=543**	

We have

$$Q_1 = value\ of\ \left(\frac{N+1}{4}\right)^{th}\ item$$

$$\therefore Q_1 = value\ of \left(\frac{543+1}{4}\right)^{th} item = value\ of\ \frac{544}{4}^{th}\ item = value\ of\ 136^{th}\ item = 40\ years$$

We have

$$Q_3 = value\ of\ 3\left(\frac{N+1}{4}\right)^{th}\ item$$

$$\therefore Q_3 = 3\left(\frac{543+1}{4}\right) = value\ of\ 3\left(\frac{544}{4}\right)^{th} item = value\ of\ 3(136)^{th} item = value\ of\ 408^{th}\ item = 60\ years$$

We know that

$$\text{Quartile deviation} = \frac{1}{2}(Q_3 - Q_1)$$

$$\therefore \text{Quartile deviation} = \frac{1}{2}(60-40) = \frac{20}{2} = 10$$

Also, Co-efficient of Quartile deviation $= \frac{Q_3 - Q_1}{Q_3 + Q_1}$

$\therefore$ Co-efficient of Quartile deviation $= \frac{60-40}{60+40} = \frac{20}{100} = 0.20$

Example: Calculate the Quartile Deviation and Co-efficient of Quartile Deviation from the following data:

Marks:	0-5	5-10	10-15	15-20	20-25	25-30
No. of Students:	4	6	8	12	7	2

Solution:

Marks	**No. of Students**	**c.f**
0-5	4	4
5-10	6	10
10-15	8	18
15-20	12	30
20-25	7	37
25-30	2	39
Total	**N=39**	

Here N=39

$$\frac{N}{4} = \frac{39}{4} = 9.75$$

$\therefore Q_1 \quad class \quad is = 5-10$

We have

$$Q_1 = l + \frac{h}{f}\left(\frac{N}{4} - C\right) = 5 + \frac{5}{6}(9.75-4) = 5 + \frac{5(5.75)}{6} = 5 + \frac{28.75}{6} = 5 + 4.79 = 9.79$$

$$\frac{3N}{4} = \frac{3(39)}{4} = 3(9.75) = 29.25$$

$\therefore Q_3 \quad class \quad is = 15-20$

We have

$$Q_3 = l + \frac{h}{f}\left(\frac{3N}{4} - C\right) = 15 + \frac{5}{12}(29.25 - 18) = 15 + \frac{5(11.25)}{12} = 15 + \frac{56.25}{12} = 15 + 4.69 = 19.69$$

We know that

$$\text{Quartile deviation} = \frac{1}{2}(Q_3 - Q_1)$$

$$\therefore \text{Quartile deviation} = \frac{1}{2}(19.69 - 9.79) = \frac{9.90}{2} = 4.95 \textit{ Marks}$$

Also, $\text{Co-efficient of Quartile deviation} = \frac{Q_3 - Q_1}{Q_3 + Q_1}$

$$\therefore \text{Co-efficient of Quartile deviation} = \frac{19.69 - 9.79}{19.69 + 9.79} = \frac{9.90}{29.48} = 0.3358 \quad \textit{Marks}$$

Merits and Demerits of Quartile Deviation

Merits

1. It is simple to comprehend and simple to calculate.
2. It isn't influenced by extreme values.
3. It can be figured for information with open end classes too.

Demerits

1. It did not depend on every one of the items. It depends on two positional qualities Q_1 and Q_3 and overlooks the extraordinary half of the things.
2. It isn't managable to further mathematical treatment.
3. It is affected by sampling fluctuations.

2.4 Mean Deviation or Average Deviation

The range and quartile deviation are not founded on all observations. They are positional measures of dispersion. They don't demonstrate any disperse of the observations from an average. The mean deviation is measure of dispersion in view of all items in a distribution.

Mean deviation is the arithmetic mean of the deviations of a series processed from any measure of central tendency; i.e., the mean, median or mode, every one of the deviations are taken as positive i.e., signs are overlooked. As per Clark and Schekade, " Average deviation is the average amount scatter of the items in a distribution from either the mean or the median, ignoring the signs of the deviations ".

We normally figure mean deviation about any of the three midpoints mean, median or mode. At times mode might be poorly characterized and as such mean deviation is computed from mean and median. Median is favoured as a decision between mean and median. But in general practice and because of wide uses of mean, the mean deviation is for the most part processed from mean. M.D can be utilized to denote mean deviation.

If $x_i,\ i=1,\ 2,\ \ldots,n$ be n observations, then mean deviation from the average A, (usually mean, mode or mode) is given by,

$$\text{Mean deviation} = \frac{1}{n}\sum_{i=1}^{n}|(x_i - A)|$$

If $x_i \mid f_i,\ i=1,\ 2,\ \ldots,n$ is the frequency distribution, then mean deviation from the average A, (usually mean, mode or mode) is given by,

$$\text{Mean deviation} = \frac{1}{N}\sum_{i=1}^{n} f_i\,|(x_i - A)|$$

Where $|(x_i - A)|$ represents the modulus or absolute value of the deviation $(x_i - A)$, where the negative sign is ignored.

Since mean deviation depends on every one of the observations, it is preferable measure of dispersion over range or quartile deviation. In any case, the progression of ignoring the signs of the deviations $(x_i - A)$ makes artificiality and renders it futile for further mathematical treatment.

Co-efficient of mean deviation: Mean deviation computed by any measure of central tendency is an absolute measure. To compare variation among various series, a relative mean deviation is required. The relative mean deviation is acquired by dividing the mean deviation by the average utilized for calculating mean deviation.

$$\text{Co-efficient of mean deviation} = \frac{Mean \quad deviation}{Mean \quad or \quad Median \quad or \quad Mode}$$

If the result is desired in percentage, then,

$$\text{Co-efficient of mean deviation} = \frac{Mean \quad deviation}{Mean \quad or \quad Median \quad or \quad Mode} X100$$

Example: Calculate mean deviation from mean and median for the following data: 100,150,200,250,360,490,500,600,671. Also, calculate coefficients of mean deviation.

Solution :

We know that

$$\text{Mean} = \frac{1}{\text{n}} \sum_{i=1}^{n} x_i$$

$$\therefore \text{Mean} = \frac{1}{9}(100+150+200+250+360+490+500+600+671) = \frac{3321}{9} = 369$$

Now arrange the data in ascending order:

100, 150, 200, 250, 360, 490, 500, 600, 671

$$\text{Median} = \text{value of} \left(\frac{\text{n}+1}{2}\right)^{th} item$$

$$\therefore \text{Median} = \text{value of} \left(\frac{9+1}{2}\right)^{th} item = \text{value of } 5^{th} \; item = 360$$

x_i	$\lvert(x_i - mean)\rvert$	$\lvert(x_i - median)\rvert$
100	$\lvert(100-369)\rvert$=269	$\lvert(100-360)\rvert$=260
150	$\lvert(150-369\rvert$=219	$\lvert(150-360\rvert$=210
200	$\lvert(200-369\rvert$=169	$\lvert(200-360\rvert$=160
250	$\lvert(250-369\rvert$=119	$\lvert(250-360\rvert$=110
360	$\lvert(360-369\rvert$=9	$\lvert(360-360\rvert$=0
490	$\lvert(490-369\rvert$=121	$\lvert(490-360\rvert$=130
500	$\lvert(500-369\rvert$=131	$\lvert(500-360\rvert$=140
600	$\lvert(600-369\rvert$=231	$\lvert(600-360\rvert$=240
671	$\lvert(671-369\rvert$=302	$\lvert(671-360\rvert$=311
	$\sum_{i=1}^{n}\lvert(x_i - mean)\rvert = 1570$	$\sum_{i=1}^{n}\lvert(x_i - median)\rvert = 1561$

$$\text{Mean deviation from mean} = \frac{1}{n}\sum_{i=1}^{n}|(x_i - mean)| = \frac{1570}{9} = 174.44$$

$$\text{Mean deviation from median} = \frac{1}{n}\sum_{i=1}^{n}|(x_i - median)| = \frac{1561}{9} = 173.44$$

$$\text{Co-efficient of mean deviation} = \frac{Mean\ \ deviation}{Mean} = \frac{174.44}{369} = 0.47$$

$$\text{Co-efficient of mean deviation} = \frac{Mean\ \ deviation}{Median} = \frac{173.44}{360} = 0.48$$

Example : Find out the mean deviation from mean and median from the following series:

Age in years	0-10	20-Oct	20-30	30-40	40-50
No. of persons	5	8	15	16	6

Also, compute co-efficient of mean deviation.

Solution: Calculation of mean and M.D. from mean

Age in years	Mid value x_i	No. of persons f_i	$d_i = x_i - a$ $= x_i - 25$	$f_i d_i$	$\lvert x_i - \bar{x}\rvert$	$f_i \lvert x_i - \bar{x}\rvert$
0-10	5	5	-20	-100	22	110
10-20	15	8	-10	-80	12	96
20-30	25	15	0	0	2	30
30-40	35	16	10	160	8	128
40-50	45	6	20	120	18	108
		$N = \sum_{i=1}^{n} f_i = 50$		$\sum_{i=1}^{n} f_i d_i = 100$		$\sum_{i=1}^{n} f_i \lvert x_i - \bar{x}\rvert = 472$

We know that

$$\bar{x} = a + \frac{1}{N}\sum_{i=1}^{n} f_i d_i = 25 + \frac{1}{50}(100) = 25 + 2 = 27$$

$$\text{Mean deviation} = \frac{1}{N}\sum_{i=1}^{n} f_i |(x_i - \bar{x})| = \frac{1}{50}(472) = 9.44 \quad marks$$

$$\text{Co-efficient of mean deviation} = \frac{Mean\ \ deviation}{Mean} = \frac{9.44}{27} = 0.35$$

Calculation of median and M.D. from median

Age in years	Mid value x_i	No. of persons f_i	c.f	$\lvert x_i - median\rvert$	$f_i \lvert x_i - median\rvert$
0-10	5	5	5	23	115
10-20	15	8	13	13	104
20-30	25	15	28	3	45
30-40	35	16	44	7	112
40-50	45	6	50	17	102
		$N = \sum_{i=1}^{n} f_i = 50$			$\sum_{i=1}^{n} f_i \lvert x_i - median\rvert = 478$

We have

$$\frac{N}{2} = \frac{50}{2} = 25$$

The cumulative frequency just greater than 25 is 28 and is corresponding class 20-30 is the median class.

$$mediam = L + \frac{h}{f}\left(\frac{N}{2} - C\right)$$

$$\therefore mediam = 20 + \frac{10}{15}\left(\frac{50}{2} - 13\right) = 20 + \frac{2}{3}(12)$$

$$= 20 + 8 = 28 \quad marks$$

$$\text{Mean deviation} = \frac{1}{N}\sum_{i=1}^{n} f_i \lvert (x_i - median)\rvert = \frac{1}{50}(478) = 9.56 \quad marks$$

$$\text{Co-efficient of mean deviation} = \frac{Mean \quad deviation}{Median} = \frac{9.56}{28} = 0.34$$

Merits and Demerits of Mean Deviation

Merits

1. It is easy to comprehend and simple to figure.
2. It is inflexibly defined.
3. It depends on all items of the series.

4. It isn't tremendously influenced by the fluctuations of sampling.
5. It is less influenced by the extreme values.
6. It is adaptable, in light of the fact that it can be computed from any average.
7. It is better proportion of examination.

Demerits

1. It's anything but an exceptionally precise measure of dispersion.
2. It isn't reasonable for further mathematical calculation.
3. It is once in a while utilized. It isn't as popular as standard deviation.
4. Algebraic positive and negative signs are ignored. It is mathematically unsound and illogical.

2.5 Standard Deviation

Karl Pearson presented the idea of standard deviation in 1893. It is the most imperative measure of dispersion and is broadly utilized in numerous statistical formulae. Standard deviation is additionally called Root-Mean Square Deviation. The reason is that it is the square-root of the mean of the squared deviation from the arithmetic mean. It gives exact result. Square of standard deviation is called Variance.

Definition

It is characterized as the positive square-base of the arithmetic mean of the square of the deviations of the given observation from their arithmetic mean. The standard deviation is signified by the Greek letter σ (sigma).

Calculation of Standard deviation-Individual Series

There are two strategies for ascertaining Standard deviation in an individual series:

a) Deviations taken from Actual mean

b) Deviation taken from Assumed mean (alternate route technique).

Deviation taken from Actual mean

This strategy is received, when the mean is an entire number.

Steps

1. Find out the actual mean of the series $(\overline{x})$
2. Find out the deviation of each value from the mean $(x_i - \overline{x})$
3. Square the deviations and take the total of squared deviations $\sum_{i=1}^{n}(x_i - \overline{x})^2$
4. Divide the total $\sum_{i=1}^{n}(x_i - \overline{x})^2$ by the number of observation n

   i.e., $\dfrac{\sum_{i=1}^{n}(x_i - \overline{x})^2}{n}$

5. The square root of $\dfrac{\sum_{i=1}^{n}(x_i - \overline{x})^2}{n}$ is standard deviation.

Thus, $\sigma = \sqrt{\dfrac{\sum_{i=1}^{n}(x_i - \overline{x})^2}{n}}$

Example : Calculate the standard deviation from the following data: 14, 22, 9, 15, 20, 17, 12, 11

Solution:

x_i	$(x_i - \overline{x}) = x_i - 15$	$(x_i - \overline{x})^2$
14	-1	1
22	7	49
9	-6	36
15	0	0
20	5	25
17	2	4
12	-3	9
11	-4	16
$\sum_{i=1}^{n} x_i = 120$		$\sum_{i=1}^{n}(x_i - \overline{x})^2 = 140$

We have

$$\bar{x} = \frac{1}{n}\sum_{i=1}^{n} x_i = \frac{1}{8}(120) = 15$$

$$\sigma = \sqrt{\frac{\sum_{i=1}^{n}(x_i - \bar{x})^2}{n}} = \sqrt{\frac{140}{8}} = \sqrt{17.5} = 4.18$$

References

1. Hague P.N. (1993). Questionnaire Design, London: Kogan Page.
2. Hogg and Tanis (2006). Probability and Statistical Inference, 7th Edition, Pearson Education, New Jersey.
3. Hildenbrand K. and Hildenbrand W. (1986). Contributions to Mathematical Economics, in K. Hildenbrand and W. Hildenbrand, On the Mean income effect: Data analysis of the U.K. Family Expenditure Survey, North Holland, New York.
4. Keyfitz N. and Beckman J.A. (1927). Demography Through Problems S-Verlag New york.
5. Marriott F.H.C. (1990). A Dictionary of Statistical Terms, 5th Edition. Longman Scientific and Technical, Harlow, Essex.
6. Danish, F. (2017). Fundamentals of Statistics: A Brief Insight, Elite Publishers, Delhi.
7. Mendenhall W. and Sincich T. (1988). A Second Course in Business Statistics: Regression Analysis 3rd Edition, San Francisco: Dellen.
8. Miller M., Irwin M. and Marylees (2006). John E. Freund's Mathematical Statistics with Applications, 7th Edition, Pearson Education, Asia.
9. Mood M., Graybill F.A. and Boes D.C. (2007). Introduction to the Theory of Statistics, 3rd Edition, Tata McGraw-Hill Pub. Co. Ltd.
10. O'Hagan A. (1994).Kendall's Advanced Theory of Statistics: Volume 2B, Bayesian Inference. Edward Arnold, London.
11. Oppenheim A.N. (1997). Questionnaire Design and Attitude Measurement, 2nd Edition, London: Heinemann.
12. Pearson K. (1914). Tables for Statisticians and Biometricians, Cambridge University Press.

13. Rowntree D. (1981). Statistics Without Tears. An Introduction for Non-mathematicians, Penguin Books, London.

14. Tarray T.A. (2016). Statistical Sample Survey Methods and Theory, Elite Publishers, Delhi.

15. Tufte G. (1983). The Visual Display of Quantitive Information, Graphics, New Heaven.

16. Tukey J.W. (1977). Exploratory Data Analysis, Addison Welsley, Reading, MA.

17. Stigler S.M. (1986). The History of Statistics: The Measurement of Uncertainty before 1900, Harvard University Press, Cambridge, MA.

18. Wahba G. (1977). Applications of statistics, in P. Krishnaiah (ed.), A survey of some smoothing problems and the method of generalized cross-validation for solving them, North Holland, Amsterdam.

19. Wahba G. (1980). Automatic Smoothing of the Log Periodogram, Journal of the American Statistical Association 75: 122-32.

20. Whittaker E.T. (1923). On a New Method Graduation, Proc. Edinburgh Math, Soc. 41: 63-75.

❑❑❑

Chapter 3

Correlation

Dr. Faizan Danish, Yasmeena Ismail* and Mir Subzar*

Department of Mathematics and Statistics, Government Degree College Shopian-192303 (J&K), India, E-mail: danishstat@gmail.com

*Division of Agricultural Statistics, SKUAST-Kashmir-1900025, E-mail: yasminjameela09@gmail.com

Correlation is a statistical technique that determines the degree of the relationship between the two different variables. Correlation is also termed as a "bivariate" statistic, with bi meaning two and variate referring to the variable or the variance. The two variables are usually a pair of scores for a person or four objects. The degree of the relationship between any two variables can vary from strong to weak or none. When a relationship between the two variables is strong, this means that if a person's or an object's score is known on one of the variables, this helps to predict the score on the second variable. We can also state it as, if a person has a high score of variable A (compared to all the other peoples' scores on A, then they are superior in score than the variable B (compared to the other peoples' scores on B). The latter would be considered as a strong positive correlation. If the correlation or relationship between the variable A and variable B is a weak one, then knowing a person's score on variable A does not help to predict their score on variable B. An interesting feature of the correlation coefficient is that it can only range from –1.00 to +1.00. Any values outside this range are invalid. Here is a graphic representation of correlation's range. Note that "r" denotes the value of correlation coefficient.

|-------------------------- |-------------------------- |---------------------- |--------------- |

-1.00 **-.50** **0** **+.50** **+1.00**

strong negative relationship **weak or none** **strong positive** **relationship**

When the correlation co-efficient r approaches positive extreme value +1.00 (or greater than r = +.50), it is an indication of **a strong positive relationship** or high degree of relationship between the two variables. A strong positive relationship means that the higher the score of a participant on one variable, the higher the score will be on the other variable. Also, if a participant attain very low score on one variable, then their score will also be low on the other variable. For instance, there is a positive correlation between years of education and wealth. Generally, the greater the number of years of education a person has, the greater their wealth. A strong correlation between these two variables also means the lower the number of years of education, the lower will be the wealth of that person. If the correlation was perfect i.e., r = +1.00, then there would not be a single exception in the entire sample to increasing years of education and increasing wealth. It would mean that there is a perfect linear relationship between the two variables. However, perfect relationships do not exist between, two variables in the real situations of statistical sampling. Thus, a strong but not a perfect relationship between the years of education and wealth in the real situations would mean that the relationship holds for most of the people in the sample but there are always some exceptions. In other words, this can be put as "some highly educated people are not wealthy, and some uneducated people are wealthy".

When the correlation co-efficient approaches r = -1.00 (or less than r = -.50), it means that there is a **strong negative relationship**. A strong negative relationship means that the higher the score of a person on one variable, the lower will be the score on the other variable. For instance, there might be a strong negative relationship between the value of gold and the Dow Jones Industrial Average. Specifically we can say that when the value of gold is high, the stock market will be lower and when the stock market is doing well, the value of gold will be lower.

A correlation coefficient that is close to r = 0.00 (note that the typical correlation coefficient is reported to two decimal places) means knowing a person's score on one variable indicates nothing about their score on the other variable. For instance, there might be a zero correlation (i.e., r = 0.00) between the number of letters in a person's last name and the number of miles they drive per day. If you know the number of letters in a last name, it tells you nothing about how many miles they drive per day. There is simply no relationship between the two variables; therefore, there is clearly a zero correlation. It should be also noted that there are no hard rules about labelling the size of a correlation co-efficient. Statisticians generally remain unexcited about a correlation until it is greater than r = 0.30 or less than r = -0.30.

The correlational statistical technique usually accompanies correlational designs. In a correlational design, the experimenter has generally a little or no control over the variables to be studied. The variables per chance got statistically analyzed long after they were initially produced or measured. Such data is called **archival**. The experimenter no longer has any experimental power to control the gathering of the data. The data has already been assembled, and the experimenter now has only statistical powers in his or her control. Cronbach (1967), an American statistician, has stated well the difference between the experimental and correlational techniques, "... the experimentalist [is] an expert puppeteer, able to keep untangled the strands to half-a-dozen independent variables. The correlational psychologist is a minimal observer of a play where Nature pulls thousand strings."

One of the prospective benefits of a correlational analysis is that *sometimes* a strong correlation between two variables may provide a guide about possible cause-effect relationships. Nevertheless, some statisticians assert a strong correlation *never* implies a cause-effect relationship. As much as correlational designs and statistical techniques are abused in this regard, the conservative statisticians' concern is understandable. It is quite thoughtful that correlational designs and techniques may allow a researcher to develop ideas about potential cause-effect relationships between variables. At that point, the researcher may conduct a controlled experiment and determine whether their cause-effect arch between two variables holds up. After completing a controlled experiment, a reaseacher can claim of cause and effect relation between two variables. Due to correlational designs and techniques that might yield clues for future controlled experimental investigations of cause-effect relationships, correlational designs and correlational statistical analyses are mostly ubiquitous in all of statistics.

Correlation : Use and Abuse

The main aim of the nature and the problem related to correlation is that, just because two variables are correlated, that doesn't mean that one variable develops the other variable. As already discussed regarding a governor who wish to give a newborn chid to every parent in his state with a classical CD or tape in order to boost the child's IQ. The governor defended his decision by refering studies, that has shown positive correlation relationship between listening classical music and intelligence. In fact, the opposition has grown to the stage where it is referred to as the Mozart Effect. The governor is assuming two wrong things,one by assuming the **causal relationship** between classical music and intelligence, which means while listening classical music the intelligence increases. However, the term 'correlation' does not permit the implication of causation.

The correlation between the two variables X and Y means that both the variables are related to each other. Thus, it is not necessary that classical music increases intelligence (x is related to y) but could be the persons who are more intelligent listen music (y is related to x). Second, the governor is taking his decision only on the basis of few correlational studies and neither he has waited for scientific repliations. It is not good to generalize that classical music raises people's intelligence, when having few correlatiobnal studies. Actually,there has been expert study of the effect that same experimenter who claimed the effect in the first place. Let's use few principles of Sagan's Baloney Detection Kit. Have the claims been verified by another source? At this point, by several researchers the effect has received little support rather than those whole claimed it forst. How does this claim fit in the world, as we know it? It does not fit very well. It would require new and undiscovered brain mechanisms. Does it seem too good to be true? Yes! it wouldn't be surprsed if just playing a Mozart CD enhanced every babies' IQ? Exactly it would provided such a surprised claim as is, we must be very skeptical. We must need higher standards of research excellence despite available research. Scientists must always be cautious. Findings must be replicated through experiments in a variety of settings with a variety of people by a variety of different researchers. Rejecting a true null hypothesis may be another uncommon illustration of Mozart Effect. What are the results of Type I error in this case? In spite of not boosting the children IQ by listening classic music, parents are exposing their children to classical music. We are witnessed to several cases of some highly educated and highly skeptical parents and grandparents who used to buy their children and grandchildren classical music toys because of having the intention of Mozart Effect. Despite knowing that there there is little probability that the Mozart Effect is real, these parents and grandparents still is a high-risk but low cost and benign consequence situation. If there's even a 1 in 10,000 chance that the Mozart Effect is real, the musical toys do not cost that much (because some kind of toy will be bought anyway) and awareness of classical music is very worst, harmless. A psychologist, Mc Burney (1996), concluded one way to counter believing in things that we like to believe is to ask to himself its consequences whether they would be true or not. For example, shouldn't everyone be listening to classical music? Wouldn't there be laws against playing any other kind of music in nurseries, kindergartens, and schools? Wouldn't we force our own children to listen to classical music? Or, do we want our children to end up dumber than the kids next door?

The Four Common Types of Correlation

1. **Pearson's r:** The type of correlation used in continuous variables that measure the strength of a relationship between two variables.
2. **Spearman's r:** That type of correlation that measures the similarity between two ordinal rankings of a single set of data.
3. **Point-Biserial r**:This type of correlation measures the strength of a relationship between one dichotomous variable (a two-level-only variable like gender)and one continuous variable, and
4. **Phi (δ) Correlation:** A measure of the strength of a relationship between two dichotomous variables.

The Pearson Product-Moment Correlation Coefficient

One of the most common used type of correlation is the **Pearson Product-Moment Correlation Coefficient**, which measures the linear degree of association between two continuous variables. For example, gold prices are continuous because they can range along a line scale from about $300 to about $500 an ounce. Gender (male or female) is not taken as continuous variable because we can assign any definite number to them like 1 or 2, a person could not be a 1.3 or a 1.7.

Pearson's co-efficient r is obtained for a sample drawn from a population. The **population value of Pearson's coefficient is called rho** (ρ), and thus, r is an estimate of ρ.

The formula for r is as follows:

$$r = \frac{N\sum xy - \left(\sum x\right)\left(\sum y\right)}{\sqrt{\left(N\sum x^2 - \left(\sum x\right)^2\right)\left(N\sum y^2 - \left(\sum y\right)^2\right)}}$$

Note: In this formula, N is equal to the number of pairs of scores and "xy is called the sum of the cross products. Let's see how the formula works in the following example:

A statistician working in any tobacco company wants to know whether heavy smoking is associated to longevity. From a study done on deceased smokers, the quantity of cigarettes (estimated on a per day for their last five years after visits with their surviving relatives) is paired with the quantity of years that they lived.

Sub.	Cigarettes	Years Lived
1	25	63
2	35	68
3	10	72
4	40	62
5	85	65
6	75	46
7	60	51
8	45	60
9	50	55

We will arbitrarily name one variable x and the other variable y. The results of Pearson's *r* will be exactly same no matter which variable is labelled x or y.

Step 1. First, obtain

$$\sum x, \sum x^2 \ (\sum x)^2, \text{ and } (\sum y)^2$$

$$\sum x = 25 + 35 + 10 + 40 + 85 + 75 + 60 + 45 + 50 = 425$$

$$\sum x^2 = 25^2 + 35^2 + 10^2 + 40^2 + 85^2 + 75^2 + 60^2 + 45^2 + 50^2 = 24{,}525$$

$$\sum y = 63 + 68 + 72 + 62 + 65 + 46 + 51 + 60 + 55 = 518$$

$$\sum y^2 = 63^2 + 68^2 + 72^2 + 62^2 + 65^2 + 46^2 + 51^2 + 60^2 + 55^2 = 33{,}188$$

$(\sum x)^2$ = HINT: square $\sum$x or $(25 + 35 + 10 + 40 + 85 + 75 + 60 + 45 + 50)^2$

$(\sum x)^2 = 425^2 = 180{,}625$

$(\sum y)^2$ = HINT: square $\sum$y or $(63 + 68 + 72 + 62 + 65 + 46 + 51 + 60 + 55)^2$

$(\sum y)^2 = 542^2 = 293{,}764$

Step 2. Obtain the sum of the cross products $(\sum \text{xy})$ by multiplying each x score by its paired y score.

$\sum xy = (25 \text{ x } 63) + (35 \text{ x } 68) + (10 \text{ x } 72) + (40 \text{ x } 62) + (85 \text{ x } 65) + (75 \text{ x } 46) +$

$(60 \text{ x } 51) + (45 \text{ x } 60) + (50 \text{ x } 55)$

$\sum xy = 1575 + 2380 + \ldots + 2750$

$\sum xy = 24{,}640$

Step 3. Obtain the value of the numerator in the *r* formula. Remember, N is equal to the number of pairs of scores. In this example, there are 9 pairs of scores.

$$N\sum xy - \left(\sum x\right)\left(\sum y\right)$$

(9) (24,640) - (425) (518)

221,760 - 230,350 = -8,590

Step 4. Obtain the value of the denominator in the r formula. Remember that the square root is obtained after all other denominator values have been computed, simplified, and reduced to a single number.

$$= \sqrt{\left(N\sum x^2 - \left(\sum x\right)^2\right)\left(N\sum y^2 - \left(\sum y\right)^2\right)}$$

$$= \frac{}{\sqrt{\left(9(24525)-(425)^2\right)\left(9(33188)-(542)^2\right)}}$$

$= 14,057.482$

Step 5. Divide the result of Step 3 by the result of Step 4.

$$r = \frac{-8590}{14057.482}$$

r = -.6111

r = -.61

It is to be noted that the Pearson r is usually taken two digits after decimal places. Thus, r = -.61 indicates the strong association between smoking and longevity, which interprets that more the quantity of cigarettes smoked in the past five years, the lower the number of years lived. And the less the quantity of cigarettes, the more the quantity of years lived. Note that the above result doesn't mean that heavy smoking is responsible to live less. It might, in any case, give pieces of information as to additionally explore thoughts for experiments. For this situation, an analysis may be set up (perhaps with animals) with an experimental group and a control group to decide, if cigarette smoking really has a causal association with early grimness.

Testing for the Significance of a Correlation Co-efficient

A correlation coefficient may be tested to determine whether the co-efficient significantly differs from zero. The value r is obtained on a sample. The value rho (└) is the population's correlation coefficient. It is hoped that r closely approximates rho. The null and alternative hypotheses are as follows:

Ho: $\rho = 0$

Ha: $\rho \neq 0$

The value of r and the number of pairs of scores are converted through a formula into a distribution (similar to the z distribution) called the ***t* distribution** (in Appendix B, page 276 of the required text, *Statistics: A Gentle Introduction*). The t formula can only be used to test whether r is equal to zero.It can't be utilized to test to see whether r may be equivalent to some number other than zero. It is additionally imperative to take note of that the t distribution might be utilized to test other types of inferential statistics. Therefore, if someone says that a *t* test is being used, it would be a legitimate question to ask "why?" The t distribution is most usually used to test whether two means are significantly different, notwithstanding, it might likewise be utilized to test the significance of the correlation coefficient. The t distribution additionally has different employments. Curiously, the t distribution moves towards becoming z distribution when the information is infinite, however, they are likewise strikingly outwardly comparable when there are just a few hundred numbers in the set of information.

The *t* test formula in order to test the null hypothesis for a correlation coefficient is:

$$t = \frac{r}{\sqrt{\frac{1-r^2}{N-2}}}$$

In the previous example on smoking, the research question was whether heavy smoking was related to longevity. A r = -.61 was obtained which meant that there was a strong negative relationship between smoking and longevity. In order to test whether this obtained r significantly differs from zero, the *t* formula is used.

$$t = \frac{r}{\sqrt{\frac{1-r^2}{N-2}}}$$

where N = the number of pairs of scores.

Obtaining the Critical Values of the *t* Distribution

We will now obtain the **critical values** of *t*, which is obtained from the *t* distribution in Appendix B.

Step 1. Choose a one-tailed or two-tailed test of significance. The alternative hypothesis establishes whether we will use a one-tailed or two-tailed significance test. If the alternative hypothesis is non-directional, as is the case in most studies, then a two-tailed test of significance is required.

Step 2. Choose the level of significance. The conventional level of significance is p = .05. Only in rare circumstances would one ever depart from p = .05 as a starting point.

Step 3. Determine the degrees of freedom (*df*). The *df* is an advanced statistical concept related to sampling. We will keep things simple. The formula for this *t* test statistic is $df = N - 2$, where N is the number of pairs of scores. In this example, there were 9 pairs of scores so $df = 9 - 2$ or $df = 7$.

Step 4. Determine whether the t from the formula (called the derived t) exceeds the tabled critical values from the t distribution. For a two-tailed test of significance at $p = .05$ with $df = 7$, let us assume the critical values of *t* are $t = +2.365$ and $t = -2.365$. If the derived *t* is greater than $t = +2.365$ or less than $t = -2.365$, then the null hypothesis will be rejected. As above, the derived $t = -2.042$ is not less than $t = -2.365$, thus, the null hypothesis is accepted, and it will be concluded that r = -.61 indicates a non-significant relationship. Interestingly, although the strength of the association was strong (*r*-.61), the test of significance shows that the obtained relationship was likely due to chance or there was higher than 5 chances out of 100 than the relationship was due to chance.

In a research paper, the results might be reported as follows:

"There was a strong negative correlation found between smoking and longevity although the correlation was not statistically significant, $r(7) = -.61$, $p > .05$. Note that the degrees of freedom appear in the parentheses to the right of *r*.

If the Null Hypothesis is Rejected

Keep in mind likewise that if the null hypothesis is rejected, the experimenter would report the most reduced p level conceivable. All things considered, the determined t would be contrasted with the basic estimations of t at .01 and .001. In the event that the determined t surpassed the critical values at both .01 and .001, at that point the experimenter would report the r as significant at $p < .001$.

Representing the Pearson Correlation Graphically: The Scatterplot

A scatterplot is a realistic introduction of the sets of scores engaged with a correlation co-efficient. It is additionally some of the time called a bivariate distribution. Give us a chance to create a scatter plot of the case on page 161 of the quantity of cigarettes smoked and life-span. Keeping in mind the end goal to develop a scatterplot, get ready to chart every one of the factors along one of the diagram's axes. At last, it doesn't make a difference which variable is decided for which axis, similarly as long as you set up every axis for one of the factors (see Figure 6.1).

The request in which the sets are diagrammed isn't essential. In this manner, it doesn't make a difference on the off chance that you start with the last match of scores or the principal combine of scores. Be that as it may, each match of scores is imperative, so make sure to plot every member's score on one variable with their relating score on the other variable.

For instance, in the event that we started with the member who smoked a normal of 85 cigarettes for every day and lived 65 years, we would first find the 85 cigarettes esteem on the cigarette axis and draw a flat line over the diagram along this value. Next, find this longevity score on the longevity axis and draw a line vertically up the graph. The crossing point of the two lines is the realistic portrayal of the match of qualities in numerical two-dimensional space.

The Assumption of Homoscedasticity

Homoscedasticity is an another assumption of the correlation co-efficient and this assumption is met only if we have the equal distances from the points to the line is relatively equal all along the line. In case the assumption is not followed, it is the case of **Heteroscedasticity**, and a graphic representation is presented in Figure 6.5.

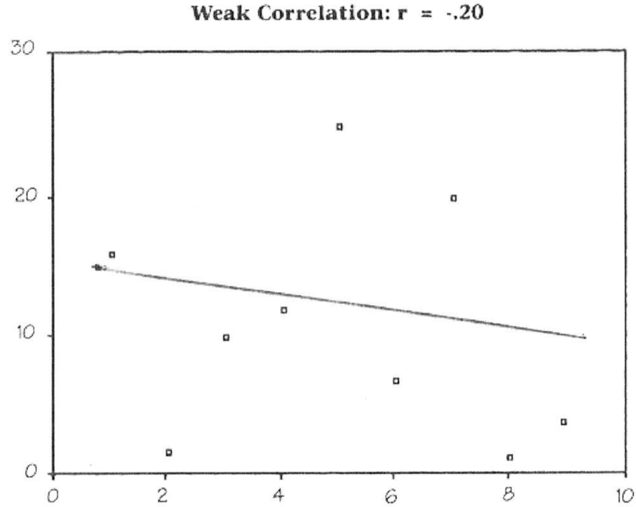

Figure 6.5

The effects of violating the assumption of homoscedasticity are the same as violating the assumption of linearity, and that is, the value of *r* is more likely to underestimate the population value of *r*.

The impacts of disregarding the supposition of homoscedasticity are the same as damaging the assumption of linearity, and that is, the estimation of r will probably to underestimate the population value of r.

The Co-efficient of Determination: How Much One Variable Accounts for Variation in Another

Variable: The Interpretation of r^2

Another utilization of the correlation coefficient is its squared esteemation r^2 is known as the co-efficient of determination, and it has two imperative understandings. To start with, it clarifies the extent of change in one variable represented by the other variable. For instance, if the relationship between two factors A and B is r = 0 .25, at that point $r^2 = (0.25)(0.25) = 0.0625$. Along these lines, the variable A clarifies roughly 6% of the variety in factor B. Another method for taking a look at r in this illustration, is say that 6% of the variation in factor B can be disclosed by its relationship to variable An, and 94% of the aggregate variation between factors A and B stays unexplained. The previous change is called shared variance and the latter variance is called **uncommon, unshared**, or **unexplained variance**.

A second understanding of r^2 is its utilization as a measure of length between at least two r values. For instance, if given a r = 0.25 and r = 0.50, obviously the second r-value is twice as awesome as the first r value. Be that as it may, the r^2 value are 6.25% and 25% respectively. In this way, r = 0.50 really explains four-fold the amount of change as much as does an r = 0.25. The good here is to be cautious when translating and looking at r values, especially smaller values of r, where any $r < 0.31$ will clarify under 10% of the change in two factors.

A similar causality alert that was connected to the interpretation of the simple correlation co-efficient r is connected to the interpretation of r^2. The co-efficient of determination does not give a clarification to the observed relationship or nor does it represent the explanation behind the relationship. The co-efficient of determination basically gives another method for survey the relationship between two factors.

Quirks in the Interpretation of Significant and Non-significant Correlation Co-efficients

There are some genuine quirks in the translations of the connection co-efficients. We have seen one as of now in our example on smoking and life-span. The understanding of the quality of the correlation between these two factors was really independent of the significance testing. The inferred r = 0-.61 showed that there was a modestly strong negative correlation amongst smoking and life-span. In any case, we found that this relationship was not statistically significant at $p < 0.05$. The statistical peculiarity is that significance tests are artifactually dependent upon sample size: Larger example sizes will more probable create significance than littler samples. In our case, N = 9 was an extraordinarily little sample, therefore, regardless of a tolerably solid r-value, we were not ready to dismiss the null hypothesis. On the off chance that we had 30 sets of scores, we would have acquired significance.

At this point, the co-efficient of determination helps us to interpret this sample size quirk. When we square $r = -0.61$, we obtain 0.37 or 37% of the variance is shared by the two variables. Now, the co-efficient of determination causes us to decipher this sample size quirk. When we square r = - 0.61, we acquire 0.37 or 37% of the variation is shared by the two factors. This is anything but a little amount of variance, things being what they are. In a 1988 preference examine, it was discovered that r = 0.06 was significant for the relationship between hostile to Semitism and bias against smoking. How could the last r be statistically significant when our r = -0.61 was most certainly not? In the bias examination there were 5,977 sets of scores. When we square r = 0.06, we acquire $r^2 = 0.0036$ or around 33% of one per cent of the aggregate change is clarified difference! This likewise demonstrates well more than 99% of the variance stays unexplained between the two factors. In these two illustrations, the co-efficient of determination has helped gather up a portion of the disarray in the interpretation of significance or an absence of significance.

References

1. Aldrich, John (1995). "Correlations Genuine and Spurious in Pearson and Yule". Statistical Science 10 (4): 364–376. doi:10.1214/ss/1177009870. JSTOR 2246135.

2. Bland J.M., Altman DG. 1996 Measurement error and correlation co-efficients. BMJ; 313: 41-2. Available from: http://ccg.cochrane.org/basic-statistics
3. Cohen, J., Cohen P., West, S.G., & Aiken, L.S. (2002). Applied multiple regression/correlation analysis for the behavioural sciences (3rd ed.). Psychology Press. ISBN 0-8058-2223-2.
4. Fisher R.A. (1915). Frequency Distribution of the Values of the Correlation Co-efficient in Samples from an Indefinitely Large Population. Biometrika, Vol. 10, No. 4 (May, 1915), pp. 507-521.
5. Francis, D.P.; Coats A.J.; Gibson D. (1999). "How high can a correlation co-efficient be?". Int J Cardiol 69(2): 185–199. doi:10.1016/S0167-5273(99)00028-5.
6. Hazewinkel, Michiel, ed. (2001), "Correlation (in statistics)", Encyclopedia of Mathematics, Springer, ISBN 978-1-55608-010-4
7. Kendall, M.G. (1955). "Rank Correlation Methods", Charles Griffin & Co.
8. Nichols D.P. (1998). Choosing An Intraclass Correlation Co-efficient From SPSS Keywords, Number 67, Available online at: http://support.spss.com/ProductsExt/SPSS/Documentation/Statistics/articles/whichicc.htm [accessed 21/04/2010 08:30]
9. Rodgers, J.L.; Nicewander, W.A. (1988). "Thirteen ways to look at the correlation co-efficient". The American Statistician 42 (1): 59–66. doi:10.1080/00031305.1988.10475524. JSTOR 2685263.
10. Rovine M.J. von Eye A. (1997). A 14th Way to Look at a Correlation Co-efficient: Correlation as the Proportion of Matches. The American Statistician, Vol. 51, No. 1 (Feb., 1997), 42-46 Stable URL: http://www.jstor.org/stable/2684692
11. Snedecor G. Cochran W. (1989). (8th ed.) Statistical Methods. Blackwell
12. Székely, G.J. Rizzo; Bakirov, N.K. (2007). "Measuring and testing independence by correlation of distances". Annals of Statistics 35 (6): 2769–2794. doi:10.1214/009053607000000505.
13. Berman, J.S., & Kenny, D.A. (1976). Correlational bias in observer ratings. Journal of Personality and Social Psychology, 34: 263-273.
14. Brewer, M.B., Campbell, D.T., & Crano, W.D. (1970). Testing a single factor model as an alternative to the misuse of partial correlations in hypothesis-testing research. Sociometry, 33: 1-11.

15. Cohen, J. and Cohen, P. (1975). Applied multiple regression/ correlation analysis for the behavioural sciences. Hillsdale, NJ: Lawrence Erlbaum.

16. Cudeck, R. (1989). The analysis of correlation matrices using covariance structure models, Psychological Bulletin, 105: 317-327.

17. Goldberger, A.S. (1964). Econometric theory, New York: Wiley.

18. Hays, W.L. (1963). Statistics. New York: Holt, Rinehart, & Winston.

19. Snedecor, G.W., & Cochran, W.G. (1967). Statistical methods, 6th ed. Ames, IA: Iowa State University.

20. Wright, S. (1921). Correlation and causation, Journal of Agricultural Research, 20: 557-585.

❑❑❑

Chapter 4

Regression Analysis

Firdoos Yousuf

Department of Statistics and Operations Research, Aligarh Muslim University, Aligarh-202002, India

4.1 Introduction

Regression analysis is a statistical method for studying linear relationships. Much of applied econometric analysis begins with the following idea: y and x are two variables, signifying some population, and we are attentive in "explaining y in terms of x," or in "reviewing how y varies with changes in x." We talk over some examples in this section, including:is soyabean crop yield and is amount of nourishment; is hourly wageand is years of education; and is a civic crime rate and is number of police officers.

Regression analysis has frequent areas of applications. A partial list would include economics, finance, business, law, meteorology, medicine, biology, chemistry, engineering, physics, education, sports, history, sociology, and psychology. Regression analysis is erudite most effectively by analysing data that are of direct interest to the bibliophile. Regression analysis is one of the most widely used statistical tools because it provides simple methods for launching a functional association among variables. It has widespread applications in many subject areas.

In text down a model that will "explain in terms of,we must defy three issues. First, since there is not once an exact relationship between two variables, how do we allow for other factors to affect ? Second, what is the efficient relationship betweenand ? And third, how can we be assured we are taking a ceteris paribus relationship betweenand ?

We can determine these uncertainties by writing down an equation relating to. A simple equation is

$$y = \beta_0 + \beta_1 x + u \qquad [1.1]$$

Equation (1.1), which is presumed to hold in the population of concern, defines the simple linear regression model. It is also called the two-variable linear regression model or bivariate linear regression model because it bears the two variables x and y. We now preserve the meaning of each of the amounts in (1.1).

When connected by (1.1), the variables and have some unlike names used interchangeably, as follows: is called the dependent variable, the response variable, the predicted variable, or the regress and; is called the independent variable, the explanatory variable, the control variable, the predictor variable, or the regressor. The terms "dependent variable"and "independent variable" are often used in econometrics. But be aware that the label "independent" here does not mention to the statistical notion of independence between random variables.

The terms "explained" and "explanatory" variables are possibly the most descriptive.

"Response" and "control" are used frequently in the experimental disciplines, where the variable is under the experimenter's control. We will not use the terms "predicted variable"and "predictor," even though you sometimes see these in tenders that are morally about prediction and not causality.

The variable, called the error term or disturbance in the relationship, signifies factors other than that affect . A simple regression analysis efficiently gives all factors affecting other than as being unnoticed.You can usefully think of as upended for "unobserved."

Equation (1.1) also talks the issue of the functional relationship between and . If the additional factors are held fixed, so that the change in is zero, then has a linear effect on :

$$\Delta y = \beta_1 \Delta x \qquad if \quad \Delta u = 0 \qquad [1.2]$$

Thus, the variation in y is simply β_1 multiplied by the change in . This means that β_1 is the slope parameter in the relationship between y and x, form the other factors in fixed; it is of chief interest in applied economics. The intercept β_0 parameter, occasionally called the constant term, also has its uses, even though it is rarely central to an scrutiny. The linear regression model, classically assessed by ordinary least squares (OLS), is the pillar of applied econometrics. The model is

$$y_i = x_i^T \beta + \varepsilon_i, \qquad i = 1, \ldots, n,$$

Or, in matrix form,

$$y = X\beta + \varepsilon,$$

where in an nx1 vector covering the dependent variable, x_1 is the (column) vector of covariates for observation i-thus $X = (x_1, \ldots, x_n)^T$ is the $n \times k$ model matrix (whose columns contain the regressors) and β is a $k \times 1$ vector of regression co-efficients. Moreover, ε is the $n \times 1$ vector of disturbances (or error terms). Assumptions on the error terms depend on the context. For cross sections, $E(\varepsilon \mid X = 0)$ (exogeneity) and $Var(\varepsilon \mid X) = \sigma^2 I$ (conditional homoskedasticity and lack of correlation) are common. Though, for time series data, exogeneity is too strong an assumption, and it is commonly replaced by pre-determinedness; i.e., $E(\varepsilon_j \mid x_i) = 0,\ i, \leq j$.

We undertake that readers are aware with the basics of the linear regression model, say at the level of Baltagi (2002) or Greene (2003). To fix notation, let $\hat{\beta} = (\hat{X}X)^{-1} X^T y$ denote the acquainted OLS estimator of β. The corresponding fitted values are $\hat{y} = X\hat{\beta}$, the residuals are $\hat{\varepsilon} = y - \hat{y}$, and the residual sum of squares (RSS) is $\hat{\varepsilon}^T \hat{\varepsilon}$.

In R, models are classically fitted by calling a model fitting function, in this case lm(), with a "formula" object labelling the model and a "data.frame" object having the variables used in the formula. Most fitting functions, including lm(), take further opinions, providing a more comprehensive explanation of the model or control parameters for the fitting procedure. By considering such further arguments in ..., a perfect call looks like fn=lm(formula, data, ...) recurring a fitted-model object, here stored in fn. This fitted model can consequently be printed, or summarized, fitted values and residuals can be mined on new data computed. Approaches for suitable generic functions such as summary(), residuals(), or predict(), etc., are presented for all standard models.

Many models can be seen as delays of the linear regression model. The R function lm() for fitting linear models is a perfect fitting function.The Pearson correlation co-efficient (ρ), is a population parameter that trials the degree of relationship between two variables. It is a usual parameter for a distribution called the bivariate normal distribution. Fleetingly, the bivariate normal distribution is a probability distribution for X and Y that has normal distributions for both X and Y and a superior form for the density function for the variable braces. This form permits for positive or negative dependence between X and Y.Many biomedical illustrations call for the use of correlation co-efficients. A surgeon might want to know whether there is an connotation between total serum cholesterol values and triglycerides. For example, researchers have studied the correlation between Center for Epidemiologic Studies Depression (CESD) scores (a measure of depressive symptoms) and stressful life events measures. Regression analysis is very closely related to linear correlation analysis. In fact, we will study that the formulations for correlation co-efficients and the slope of a regression line are similar and functionally connected. Thus, far we have distributed with bivariate examples, but linear regression can extend to more than one predictor variable. The linearity requisite in the model is for the regression co-efficients and not for the predictor variables. We will deliver more information on multiple regression in Section 1.2.

Detectives use regression analysis very broadly in the biomedical sciences. As noted previously, the researchers use an independent variable to expect a dependent variable. For example, regression analysis may be used to consider a dose-response association for a drug managed to laboratory animals. The drug dose would be deliberated the independent variable, and the response chosen would be the dependent variable. A dose-response association is a type of relationship in which growing doses of a material produce growing biological responses; e.g., the association between number of cigarettes consumed and incidence of lung cancer is reflected to be a dose-response association. The simple regression model can be used to re-vision the association between two variables. For reasons we will see, the simple regression model has confines as a general tool for realistic analysis. However, it is occasionally appropriate as an empirical tool. Learning how to understand the simple regression model is good practice for reading multiple regressions, which we will do in subsequent sections.

4.2 Assumptions of the Linear Regression Model

The linear regression model comprises of a set of assumptions about how a data set will be created by an underlying "data producing process." The theory will identify a deterministic association between the dependent variable and the independent variable. The assumptions that designate the form of the model and associations among its parts and imply applicable estimation and inference measures are listed in Table 1.1.

4.2.1 Linearity of the Regression Model

Let the column vector x_k be the n observations on variable x_k, $k = 1,\ldots,K$, and gather these data in an $n \times K$ data matrix, X. In most circumstances, the first column of X is assumed to be a column of 1s so that β_1 is the constant term in the model. Let y be the n observations $y_1,\ldots,y_n$, and let ε be the column vector having the n disturbances.

Table 1. Assumptions of the Linear Regression Model

A1. **Linearity :** $y_i = x_{i1}\beta_1 + x_{i2}\beta_2 + \ldots + x_{ik}\beta_k + \varepsilon_i$. The model stipulates a linear association between and $x_1,\ldots,x_k$.

A2. **Full rank :** There is no exact linear association among any of the independent variables in the model. This assumption will be required for estimation of the parameters of the model.

A3. **Exogeneity of the independent variables :** $E\left[\varepsilon_i \mid x_{j1}, x_{j2},\ldots,x_{jk}\right] = 0$ This states that the expected value of the disturbance at observation i in the sample is not a function of the independent variables observed at any observation, including this one. This means that the independent variables will not convey useful information for prediction of ε_i.

A4. **Homoscedasticity and non-autocorrelation :** Each disturbance, εi has the same finite variance, σ^2, and is uncorrelated with every other disturbance, ε_j. This assumption bounds the generalization of the model, and we will want to inspect how to relax it in the chapters to follow:

A5. **Data generation:** The data in $\left(x_{j1},\ldots,x_{jk}\right)$ may be any mixture of constants and random variables. The crucial elements for present determinations are the strict mean independence assumption A3 and the implicit variance independence assumption in A4. Analysis will be done conditionally on the observed X, so whether the elements in X are fixed constants or random draws from a stochastic process will not influence the results. In future, more advanced treatments, we will want to be more precise about the possible association between ε_i and x_j.

A6. **Normal distribution :** The disturbances are normally distributed. Once again, this is a suitability that we will dispense with after some analysis of its insinuations.

4.3 Non-linear Regression Models

A non-linear regression model is one for which the first order circumstances for least squares estimation of the parameters are non-linear functions of the parameters. The common form of the non-linear regression model is

$$y_i = h(x_i, \beta) + \varepsilon_i.$$

The linear model is visibly a special case. Furthermore, some models that perform to be non-linear, such as:

$$y = e^{\beta_1} x_1^{\beta_1} x_2^{\beta_2} e^{\varepsilon}$$

become linear after an alteration, in this case after taking logarithms.

4.3.1 Assumptions of the Non-linear Regression Model

We shall need a fairly more formal meaning of a non-linear regression model. Sufficient for our drives will be the following, which include the linear model as the special case noted earlier. We adopt that there is an fundamental probability distribution, or data generating process (DGP) for the visible and a true parameter vector, β, which is a characteristic of that DGP. The following are the assumptions of the non-linear regression model:

1. **Functional form :** The conditional mean function for given is:

   $E[y_i \mid x_i] = h(x_i, \beta), \quad i = 1, \dots, n,$

   where $h(x_i, \beta)$ is a continuously differentiable function of β.

2. **Identifiability of the model parameters :** The parameter vector in the model is recognized (estimable), if there is no non-zero parameter $\beta^0 \neq \beta$ such that $h(x_i, \beta^0) = h(x_i, \beta)$ for all x_i. In the linear model, this was the full rank assumption, but the simple absenteeism of "multicollinearity" among the variables in x is not sufficient to harvest this condition in the non-linear regression model.

3. **Zero mean of the disturbance :** It follows from Assumption 1 that we may write $y_i = h(x_i, \beta) + \varepsilon_i$, where $E[\varepsilon_i \mid h(x_i, \beta)] = 0$. This states that the disturbance at observation i is uncorrelated with the conditional mean function for all observations in the sample. This is not fairly the same as assuming that the disturbances and the exogenous variables are uncorrelated, which is the familiar assumption, however.

4. **Homoscedasticity and non-autocorrelation :** As in the linear model, we adopt conditional homoscedasticity,

   $E[\varepsilon_i^2 \mid h(x_j, \beta), \quad j = 1, \dots, n] = \sigma^2$, a finite constant

   and non- auto correlation

   $E[\varepsilon_i \varepsilon_j \mid h(x_i, \beta), h(x_j, \beta), j = 1, \dots, n] = 0 \quad \text{for all} \quad j \neq i.$

5. **Data generating process :** The data generating process for **x***i* is assumed to be a well-mannered population such that first and second moments of the data can be presumed to converge to fixed, finite population complements. The crucial assumption is that the process generating x_i is stringently exogenous to that generating ε_i. The data on x_i are assumed to be good.

6. **Underlying probability model :** There is a vague probability distribution generating ε_i. At this point, we assume only that this process harvests a sample of uncorrelated, identically (marginally) distributed random variables ε_i with mean zero and variance σ^2 conditioned $h(x_i, \beta)$. Thus, at this point, our

statement of the model is semi-parametric. We will not be presumptuous any particular distribution for ε_i. The conditional moment assumptions in 3 and 4 will be adequate for the results in this chapter. This will allow us to be more precise about certain test statistics and, in addition, allow some generalizations of the regression model. The assumption is not necessary here.

4.4 Multiple Linear Regression

Multiple linear regression is an extension (generalization) of simple linear regression. One can correspondingly think of simple regression as an exceptional case of multiple regression because all simple regression results can be obtained using the multiple regression. In economics, maximum regression analyses embrace more than a single regressor often there are regressors of a superior type, characteristically referred to as "dummy variables" in econometrics, which are used for coding categorical variables. Additionally, it is often also obligatory to transmute regressors or dependent variables. To exemplify how to deal with this in R software, we consider a usual task in labour economics, estimate of a wage equation in semi-logarithmic form. Here, we hire the CPS1988 data frame placed in the March 1988 Current Population Survey (CPS) by the US Census Bureau and analyzed by Bierens and Ginther (2001). In this data frame, there are 28155 observations on seven variables. This is the cross-section data originating from the March 1988 Current Population Survey by the US Census Bureau. Wages are emptied by the deflator of personal consumption expenditures for 1992. A summary of the data set can be obtained as usual:

```
data("CPS1988",package="AER")
head(CPS1988)
```

Wage	Education	Experience	Ethnicity	Smsaregion	Part time
1.	354.94	7	45	caucyes northeast	no
2.	123.46	12	1	caucyes northeast	yes
3.	370.37	9	9	caucyes northeast	no
4.	754.94	11	46	caucyes northeast	no
5.	593.54	12	36	caucyes northeast	no
6.	377.23	16	22	caucyes northeast	no

Summary (CPS, 1988)

wage education experience ethnicity smsaregion part time

Min.: 50.05 Min.: 0.00 Min.:-4.0 cauc:25923no:7223 northeast:6441 no:25631

1st Qu.: 308.64 1st Qu.: 12.00 1st Qu.: 8.0 afam: 2232 yes:20932 midwest:6862 yes:2524

Median: 522.32 Media : 12.00 Median :16.0 south :8760

Mean : 603.73 Mean:13.07 Mean :18.2 west :6091

3rd Qu :783.48 3rd Qu.: 15.00 3rd Qu.:27.0

Max. : 18777.20 Max. :18.00Max.:63.0

The model of interest is

$$\log(wage) = \beta_1 + \beta_2 \exp erience + \beta_3 \exp erience^2 + \beta_4 eduction + \beta_5 ethnicity + \varepsilon$$

Which can be fitted in R using

R>M0<-lm(log(wage)~experience + I (experience^2) + education + ethnicity, data = CPS, 1988)

The formula in the lm() call takes into account the semi-logarithmic form and also stipulates the squared regressor experience^2. It has to be insulated by I() so that the operator ^ has its original arithmetic meaning.

Multiple regression analysis is more agreeable to ceteris paribus analysis because it allows us to openly control for many other factors that instantaneously affect the dependent variable. This is important both for testing economic philosophies and for appraising policy effects when we must rely on non-experimental data. Because multiple regression models can lodge many explanatory variables that may be correlated, we can hope to infer destiny in cases where simple regression analysis would be ambiguous.

Naturally, if we add more factors to our model that are useful for explaining , then more of the variation in can be clarified. Thus, multiple regression analysis can be used to construct better models for predicting the dependent variable.

An additional benefit of multiple regression analysis is that it can integrate fairly general functional form associations. In the simple regression model, only one function of a single explanatory variable can be seen in the equation. As we will see, the multiple regression model allows for much more litheness.

The multiple regression model is still the most extensively used vehicle for empirical analysis in economics and other social sciences. Likewise, the method of ordinary least squares is popularly used for estimating the parameters of the multiple regression model.

4.5 Motivation for Multiple Regressions

We begin with some simple examples to show how multiple regression analysis can be used to solve problems that cannot be solved by simple regression.

The first example is a simple variation of the wage equation presented in previous section for obtaining the effect of education on hourly wage:

$$wage = \beta_0 + \beta_1 edu + \beta_2 \exp er + u, \qquad [1.3]$$

where exper is years of labour market experience. Thus, wage is resolute by the two explanatory or independent variables, education and experience, and by other unobserved factors, which are checked in u. We are still chiefly interested in the effect of edu on wage, holding fixed all other factors distressing wage; that is, we are attentive in the parameter β_1 .

Compared with a simple regression analysis relating wage to edu, equation (1.3) efficiently takes exper out of the error term and puts it obviously in the equation. Because exper seems in the equation, its co-efficient, β_2, dealings the ceteris paribus effect of experon wage, which is also of some interest.

Not astonishingly, just as with simple regression, we will have to make expectations about how u in (3.1) is related to the independent variables, educand exper. However, as we will see in Section 1.2, around is one thing of which we can be confident: because (1.3) contains experience plainly, we will be able to measure the effect of education on wage, holding experience fixed. In a simple regression analysis which puts exper in the error term we would have to undertake that experience is uncorrelated with education, a tenuous assumption.

In the previous similar example, we have shown how noticeable factors other than the variable of primary interest [educ in equation (1.3) can be included in a regression model. Generally, we can write a model with two independent variables as:

$$y = \beta_0 + \beta_1 x_1 + \beta_2 x_2 + u, \quad [1.4]$$

Where,

β_0 is the intercept.

β_2 measures the change in y with respect to x_1, holding other factors fixed.

β_3 measures the change in y with respect to x_2, holding other factors fixed.

Multiple regression analysis is also beneficial for generalizing functional associations between variables. As an example, suppose family consumption (*cons*) is a quadratic function of family income (*inc*):

$$cons = \beta_0 + \beta_1 inc + \beta_2 inc^2 + u, \quad [1.5]$$

where u contains other factors affecting consumption. In this model, consumption depends on only one observed factor, income; so it might seem that it can be controlled in a simple regression framework. But the model falls outside simple regression because it contains two functions of income, inc and inc^2 (and therefore three parameters, β_0, β_1,and β_2). Nevertheless, the consumption function is easily written as a regression model with two independent variables by letting $x_1 = inc$ and $x_2 = inc^2$.

In other words, the Marginal effect of income on consumption be contingent on β_2 as well as on β_1 and the level of income. This example shows that, in any particular application, the descriptions of the independent variables are crucial. But for the theoretical development of multiple regression, we can be imprecise about such details.

In the model with two independent variables, the key assumption about how is related to and is

$$E(u \mid x_1, x_2) = 0. \quad [1.6]$$

It means that, for any values of and in the population, the average of the unobserved aspects is equal to zero. As with simple regression, the significant part of the assumption is that the predictable value of u is the same for all combinations of x_1 and x_2 ; that this common value is zero is no assumption at all as long as the intercept β_0 is included in the model (see Section 1.1).

On the one hand, all of the estimates, tests, and other synopses are computed as if the model and its assumptions were accurate, but on the other, there exist various regression methods apart from OLS, some of which are more appropriate in certain applications.

Here, we discuss several approaches to validating linear regression models:

- A popular approach (on occasion somewhat misleading) compares various statistics computed for the full data set with those obtained from deleting single observations. This is known as regression diagnostics.
- In econometrics, diagnostic tests have played a protuberant role since about 1980. The most important alternative hypotheses are heteroskedasticity, auto correlation,and misspecification of the functional form.
- Also, the impenetrable disturbance edifices typically present in observational data have led to the development of "robust" covariance matrix estimators, a number of which have become available during the last 20years.

Additionally, resilient (in the sense of resistance to outliers and unusual observations) regression practices are often quite helpful, although such methods do not appear to be as widely known among econometricians as they deserve to be. We also include a brief introduction to quantile regression, a method that has been receiving increasing interest in applied econometrics, especially in labour economics.

4.6 Summary and Conclusions

This chapter has enclosed the linear regression model, the basic platform for model building in econometrics. The assumptions of the conventional regression model are abridged in table 1.1, which shows the two variable case. We prolonged the regression model to a form that permits non-linearity in the parameters in the regression function. The results for interpretation, estimation, and hypothesis testing are quite similar to those for the linear model. The two critical differences between the two models are, first, the more involved estimation procedures needed for the non-linear model and, second, the ambiguity of the interpretation of the coefficients in the non-linear model (because the derivatives of the regression are often non-constant, in contrast to those in the linear model).

4.7 References

1. Baltagi, B.H., S.H. Song and B.C. Jung, (2002a). A comparative study of alternative estimators for the unbalanced two-way error component regression model, Econometrics Journal 5: 480-493.
2. Baltagi, B.H., S.H. Song and B.C. Jung, (2002b). Simple LM tests for the unbalanced nested error component regression model, Econometric Reviews 21: 167–187.
3. Baltagi, B.H., S.H. Song and B.C. Jung, (2002c). LM tests for the unbalanced nested panel data regression model with serially correlated errors, Annales D'Economie et de Statistique 65:219–268.
4. Greene, W.H. (2003). *Econometric Analysis*, 5th edn. (Pearson Education, Inc., Upper Saddle River, NJ).
5. Bierens H., Ginther D.K. (2001). "Integrated Conditional Moment Testing of Quantile Regression Models." Empirical Economics, 26: 307–324.
6. Chatterjee, S. and Hadi, A.S. (1988), *Sensitivity Analysis in Linear Regression,* New York: John Wiley & Sons.
7. Chatterjee, S., Handcock, M.S., and Simonoff, J.S. (1995). *A Casebook for a First Course in Statistics and Data Analysis,* New York: John Wiley & Sons.
8. Bates, D.M. and Watts, D.G. (1988). *Non-linear Regression Analysis and Its Applications,* New York: John Wiley & Sons.

□□□

Chapter 5

Sampling Theory

Dr. Faizan Danish

Department of Mathematics and Statistics, Government Degree College Shopian (J&K) India-192303, E-mail: danishstat@gmail.com

Population

Any collection of specified group of human or non-human entities, such as objects, educational institutions, time units, geographical areas, prices of wheat is known as population or universe. A population containing a finite number of individuals, members or units is a class. A population with infinite number of members is known as infinite population. The population of trees on the whole earth is an example of infinite population. The population of concrete individuals is called as existent population, while as the collection of all possible ways in which an event can materialize as the hypothetical population. All the 400 students of 10th class of particular school is an example of existent type of population and the population of heads and tails obtained by tossing a coin on infinite number of times is an example of hypothetical population.

The population is properly defined so that there is no ambiguity as to whether a given unit belongs to the population. For example, in a survey of achievement in mathematics, a researcher will have to define the population of students by age or by grade and, if necessary, he will also specify the type of schools, the geographical area and the academic year for which the data will be collected. Inferences concerning a population cannot be drawn until the nature of the units that comprise it is clearly identified. If a population is not properly defined, a researcher does not know what units to consider, when selecting the sample.

Sample

A selected group of some elements from the totality of the population is known as the sample. It is from the study of this sample that something is known and said about the whole population. The assumption is that what is revealed about the sample will be true about the population as a whole. But it may not be true always, as it depends on the way the sample is drawn. If the sample is a replica of the population, the foregoing assumption is true. But, if it is biased, such inferences about the population cannot be true.

A biased sample is one that is selected in such a way that it yields a sample value which is much different from the true or population value. Hence, it is basic requirement for inferential research that the sample should be free from bias. In other words, it should be representative of the population. A representative sample is a sample which has all those characteristics present in the same amount or intensity in which they are found in the population. Bias in selecting a sample can be avoided and it can be made representative of the population by selecting it randomly. A random sample involves small errors in predicting population value and this error can be estimated also. Thus, the objective should always be to draw an unbiased random and representative sample.

In order to draw such a representative sample, a sample plan has to be prepared. It means a plan which, if properly executed, can guarantee that if we were to repeat a study on a number of different samples, each of a particular size drawn from a given population, our findings would not differ from the findings that we would get if the given population as a whole was studied by more than specified proportions of sample. For example, not more than 5 points in 90% of the samples, that is, out of 100 samples the sample value (estimate of value) will be correct within 5 points in 90 out of 100 samples. If the plan guarantees sufficiently well that the chances are great enough that the selected sample is representative of the population, it is known as a representative sampling plan. It ensures selecting diverse elements and making sure that these diverse elements are adequately represented in the sample.

Sampling

It is the process of selecting a sample from the population. For this purpose, the population is divided into a number of parts called sampling units. Most of the educational phenomena consist of a large number of units. It would be impracticable, if not possible to test or interview or observe each unit of the population under controlled conditions in order to arrive at principles having universal validity. Some populations are so large that their study would be expensive in terms of time, effort, money and manpower. Sampling is a process by which a relatively small number of individuals or measures of objects or events is selected and analyzed in order to find out something about the entire population from which it was selected. It helps to reduce expenditure, save time and energy, permit measurement of greater scope, and produce greater precision and accuracy.

Definitions

W.G. Cocharn

"In every branch of science, we lack the resources to study more than a fragment of the phenomena that might advance our knowledge." In this definition a 'fragment' is the sample and 'phenomena' is the population. The sample observations are applied to the phenomena, i.e., generalization.

David. S. Fox

"In the social sciences, it is not possible to collect data from every respondent relevant to our study but only from some fractional part of the respondents. The process of selecting the fractional part is called sampling." 'Sampling design' means the joint procedure of selection and estimation. Sampling should be such that error of estimation is minimum.

Taking a few observation from the population is really an economic advantage of sampling over census. For instance, let us suppose you want to know the status of yourself among all the students around and you are there about whom you have heard so much about the famous Cornell and now that you are there, Cornell. You know all the students are different, so they may have different perceptions and you wish to know all these perceptions or you want to study them in depth. In order to achieve such results, you are supposed to interview each and every student that took results in 20 days only, let us assume this particular time you are doing your research. Cornell has only 20,000 students and those who are helping are so fast at the interviewing art that together you can interview at least 10 students per person per day in addition to your 18 credit hours of course work. You need 20 days with 100 research assistants and since you are paying them minimum wage of \$5.00 per hour for ten hours (\$50.00) per person per day, you need to have \$100000.00 to finish the interviews, evaluation will not be possible for such study. You may decide to hire additional assistants to help with the analysis at another \$100000.00 and so on assuming you have that amount on your account.

Bias and Error in Sampling

A sample is expected to mirror the population from which it comes, however, there is no guarantee that any sample will be precisely representative of the population from which it comes. It may be possible that a non-desired number of untypical observations will be made like for the case of testing fuses, it may consist less or more default fuses rather than the real population proportion of faulty cases. In practice, it is rarely known when a sample is unrepresentative and should be discarded.

Sampling Error

What is the main cause of sample that would not represent the error? One of the main cause is due to sampling error. Sampling error is defined as the deviation between the sample estimate and the population estimate that are due solely to the particular units that happen to have been selected. For instance, let us suppose that a sample of 100 American women were evaluated and it was found that all found to be taller than six feet. From the result, it is obvious that the sample doesn't represent the population that leads to invalid conclusion. These cases realy happen because naturally such cases are widely distributed among the population. But it can occur. Luckily, this is a very obvious error and can be ejected very easily. One of the most dangerous error is less sampling error for which the less protection is provided by nature. One of the example through which it can be explained is like a sample in which it is overstated that the average height is only one inch or two rather than one foot which is more obvious. It is the unobvious error that is of much concern. There are two basic causes for sampling error. One is chance: That is the error that occurs just because of bad luck. This may result in untypical choices. It is possible that a number of absurd units may be selected which is unusual. For example, A study has been done related to the number of trees but when the sample was selected the two highest households were selected that were contributing (10,018 and 6345) that affects the result to a very large extent. In order to overthrow such errors, a large sample size is suggested to select. The main protection against this kind of error is to use a large enough sample. One of the other causes behind sampling is known by the sampling. Sampling bias is a tendency to favour the selection of units that have particular characteristics. Sampling bias is nothing but the response of poor sampling plan. When some units can't be included in the sample by either of the cause, the error created by this way is called non-response error. For example, let us take a hypothetical case in which a survey is conducted to study the level of stress that graduate students were going through. In order to take conclude regarding this, a questionnaire was sent to 100 randomly selected graduate through mail but unfortunately only 52 respondents responded positively and the conclusion was that students were not in stress at that time but the reality is that it was the highest time of stress for all students except for those who were busy with writing their thesis by their own efforts. The investigator went back to know the cause behind non-response of respondents and found that these were only third and fourth Ph.D students. Bias can be very costly and has to be guarded against as much as possible. One other example can be to study the average income of any community and the selection of sample is made on telephone numbers but only rich and middle class households have telephone lines that could conclude in high average income that would lead to another example, would be where you would like to know the average income of some community and you decide to use the telephone numbers to select a sample of the total population in a harmful policy decision.

Non-sampling Error (measurement error)

Non-sampling error is one other cause behind non-representative sample and this type of error can occur whether a census or a sample is being used. Like sampling error, non-sampling error may either created by participants in the statistical study during sampling plans and procedures. Inaccurate measurements due to malfunctioning instruments or poor procedures are some of the causes due to which the non-sampling error could arise. For example, consider the weights of human and if persons are supposed to record their own weights, no two answers will be of equal reliability. The people will have weighed themselves on different scales in various states of poor calibration. An individual's weight changes by several grams that may affect the exact answer. The scale reading will also cause variation with individuals undress. Therefore, the results cannot be compared unless all the individuals are weighted under same circumstances. Biased observations due to inaccurate measurement can be innocent but very devastating. A story related to a French astronomer who once developed a new theory, based on spectroscopic measurements of light emitted by a particular star and when his friends discovered that the measuring instrument is filled with or suffering from fault by presence of cigarette smoke, they didn't accept his theory. In surveys of personal characteristics, unknown errors may come from: The way by which response is fetched, the social desirability of the persons who were investigated. The aim of the work - the biases caused due to personal prejudice of the investigator.

The Interwiers effect

No two interviewers are alike and the same person may provide different answers to different interviewers. Inaccurate results can also be obtained by framing the questions in wrong manner. Individuals tend to provide false answers to particular questions. For instance, due to some reasons some people feel to look younger or older. When asked such a person reading age in years, it is not difficult for them to lie by over stating their age by many years than it is, if you asked which year they were born since it will take a few of immediate arithmetic to give a wrong date and a date of birth will definitely be more accurate.

The Respondent effect

By impressing the interviewer respondents might also give incorrect answers. Such type of errors difficult to prevent because it comes from outside on the part of respondee. An explanation of this case may be given by witnessing a case in which I was asking farmers how much maize they harvested last year (1995). In majority times, men use to lie by saying that the yield is 25 bags per acre. While comparing with the responses of the wives of the these men, their responses were all different. Choosing the right one, whenever possible I could in a tactful way verify with an older son or daughter.

Knowing the study purpose

Knowing why a study is being conducted may create incorrect responses. A classic example is the question: What is your income? In case the government organization is asking, it is obvious that a complete different figure would be provided than the actual figure provided by the respondents on the application. To overthrow such errors, one of the way is to camouflage the study's goals and make the question specific regarding the study so that no personal prejudice is included. For example, "Where are you employed?" could be followed by "What is your salary?" and "Do you have any extra jobs?" A sequence of such questions may produce more accurate information.

Induced bias

The point to be noted that either the personal bias or the collector of the data may tend to induce bias and while framing the questionnaire, the questions should be framed in such a way so that response shall be obtained even though it is inaccurate. For instance, applying fertilizers to the yield, an agronomist would expect to have more production than others. In order to protect the study from biases an expert of statistics may be desired to desighn the study.

Selecting the Sample

Vulnerability to error and its cost of a sampling procedure is directly proportional to the desirability and in preceding sections, only simple problem have been discussed with statistical studies but economy and reliability are competing ends, because, increased expenditure of resources is one of the main cause reducing error. Of the two types of statistical errors, only sampling error can be controlled by exercising care in determining the method for choosing the sample. As discussed above, bias and chance are the main causes behind sampling error and random component would always occur despite our extreme efforts to neglect that type of error and we can minimize this random error only by taking large samples. Selecting the desired sampling plan would minimize the sampling error.

Probability Sampling

In probability sampling, the units are not selected according to the wish of the researcher but the units are selected in such a way that each and every unit has an equal chance of included in sample selected for study. Such a method is also called random sampling.

Characteristics

1. Each unit in the sample has some known probability of entering the sample.
2. Weights appropriate to the probabilities are used in the analysis of the sample.
3. Its observations are used for the inferential purpose.
4. Inferential or parametric statistics are used.
5. There is risk for drawing conclusions.
6. Its representatives refer to characteristics.
7. Its comprehensiveness refers to size and area.

Techniques of probability sampling

1. Simple random sampling

In simple random sampling, each unit of the population is given an equal chance of being selected. The selection of units from the population is done in such a manner that every unit in the population has an equal chance of being chosen and the selection of any one unit is in no way tied to the selection of any other. The law of chance is allowed to operate freely in the selection. Carefully controlled conditions are created to ensure that each unit in the population has an equal chance of being included in the sample. To prevent the researcher from being the results by exercising direct control over the selection of units, several devices are employed to draw samples from population. Some techniques are as under:

a) Tossing a coin.
b) Throwing a dice.
c) Lottery method.
d) Blind folded method.
e) By using random table or 'Tippet's table'.

Merits

a) It requires a minimum knowledge of population.
b) It is free from subjectivity and free from personal error.
c) It provides appropriate data for one's purpose.
d) The observations of the sample can be used for inferential purpose.

Demerits

a) It cannot ensure representativeness of a sample.

b) It is independent of population knowledge.

c) Its inferential accuracy depends upon the size of the sample.

2. Cluster sampling

It means selecting the intact group as a whole. In it the sample units contain groups of elements (clusters) instead of individual members or items in the population. For example, a researcher lists all of the degree colleges in the city, select at random 15% of these clusters of units and uses all of the students in the selected colleges as a sample.

Merits

a) It is a good representative of the population.

b) It is an easy method.

c) It is an economical method.

d) It is highly applicable in education.

e) Its observations can be used for inferential purpose.

Demerits

a) It is not free from errors.

b) It is not comprehensive.

Stratified sampling

It is an improvement over systematic method. In this sampling, the researcher divides the heterogeneous population in small population known as strata such that the strata are homogeneous within themselves and heterogeneous between themselves and then the sample is selected from each stratum randomly. He should choose that characteristics or criterion which seems to be more relevant in his research work.

Types of stratified sampling

a) **Disproportionate sampling :** This is defined as the sampling in which the sample selected from each stratum is not proportional to its size but is based judgment and convenience. Comparing strata having different possibilities of error, such methods are suggested. It is less for determining population characteristics.

b) **Proportionate sampling :** It refers to the selection of a sample from each sampling unit that is proportionate to the size of the unit. Its advantage include representativeness with respect to variables used as the basis of classifying categories and increased chances of the comparison between strata. Lack of information on proportion of the population in each category and faulty classification are the disadvantages of this method.

c) **Optimum allocation stratified sampling :** It is representative as well as comprehensive. It refers to selecting unit from each stratum in proportion to the corresponding stratum of the population.

Merits

a) It is a good representative of the population.

b) It is an improvement over the earlier.

c) It is an objective method of sampling.

d) Observations can be used for inferential purpose.

Demerits

a) It is difficult for the researcher to decide the relevant criterion for stratification.

b) Only the criterion can be used of stratification.

c) It is costly and time consuming.

d) **Multi-stage sampling :** It is more comprehensive and representative of the population. In this type of sampling, the units selected at primary stage are inclusive but the units at second stage are sub-groups within these ultimate units to be selected which belong to one and only one group. Whenever stratification is done by the researcher, stages of population are usually available within a group or population. The individuals are selected from different stages for constituting the multi-stage sampling.

Merits

1. It is a good representative of the population.

2. It is an improvement over the earlier methods.
3. It is an objective procedure of sampling.
4. Its observations may be used for inferential purpose.

Demerits

1. It is difficult and complex.
2. It involves errors while considering the primary and secondary stages.
3. It is subjective.

Non-probability Sampling

In non-probability sampling, the units are selected at the discretion of the researcher. Such samples use human judgement in selecting units and have no theoretical basis for estimating population characteristics. If the researcher is to select a sample of 200 school students, he may exercise his own judgement based on experience for finding a given student in the sample. Such a sample is arbitrarily selected because there is good evidence that it is a representative of the total population. The evidence is always based on experience. For example, in a study of delinquency, the researcher, on the basis of his past experience, selects certain areas that have shown higher rate of delinquent acts in the past. In non-probability sampling, the cases are selected on such bases as availability and interviewer judgement. The non-probability methods are very convenient in the situations, when the sample to be selected is very small and the researcher wants to get some idea of the population characteristics in a short time.

Characteristics

1. There is no idea of population.
2. There is no probability of selecting any individual.
3. It has free distribution.
4. The observations are not used for generalization purpose.
5. Non-parametric and non-inferential statistics are used.
6. There is no risk for drawing conclusions.

Types of Non-probability Sampling

1. Quota sampling

It combines both judgement sampling and probability sampling. In it, the population is classified into several categories. The proportion of population falling into each category is decided on the basis of judgement or assumption or the previous knowledge. It is fixed to draw a quota of cases and the observer is allowed to sample as he likes. It is an arbitrary method likely to figure in municipal surveys.

Merits

1. It is an improvement over the judgement sampling.
2. It is an easy sampling technique.
3. It is most frequently used in social surveys.

Demerits

1. It is not a representative sample.
2. It is not free from errors.
3. It has the influence of regional, geographical and social factors.

2. Judgement sampling

It involves the selection of a group from the population on the basis of available information, though it should be representative of the total population. It involves the selection of a group by intuition on the basis of criterion deemed to be self-evident. Since the decision is wholly dependent upon investigator, thus this sampling is highly risky.

Merits

1. In it, knowledge of the investigator can be best used.
2. It is economical.

Demerits

1. It is not free from errors.
2. It includes uncontrolled variation.
3. In it, inferential statistics cannot be used of observations. Therefore, generalization is not possible.

3.Purposive

It is selected by some arbitrary method because it is known to be representative of the total population. It is known that it will produce well-matched groups. It picks out the sample in relation to some criterion, which are considered important for a particular study. It is appropriate in a study which lays emphasis on the control of certain specific variables.

Merits

1. It uses the best available knowledge concerning the ample subjects.
2. It gives better control to significant variables.
3. In it, sample groups data can be easily matched.
4. In it, there is homogeneity of subjects used in sample.

Demerits

1. In it, the reliability of the criterion is questionable.
2. In it, the knowledge of population is essential.
3. There may be errors in classifying sampling subjects.
4. It is usable to utilize the inferential parametric statistics.
5. It is unable to make generalization concerning total population.

Conclusion

We can conclude that using sampling saves time and money provided a suitable sampling strategy is used, appropriate sample size selected and some important considerations taken to minimize on sampling and non-sampling errors, then a sample should give valid and relative information. In case of more information, regarding sampling can be obtained in the below mentioned references and many other books related to statistics. In other words, we can conclude that researchers may choose from a variety of sampling methods. The researcher goals inform which sampling method is best for the research to be conducted. The main choice in regards to sample method choice is whether or not the researcher wants to generalize the findings from the sample to the whole of the population being studied. Being aware of possible errors due to the sample method chosen is also very important because giving possible errors within the results section allows the study to be regarded as valid. Many sample method choices are available; the researcher must choose the method that is right for the study.

References

1. Danish, F. (2017). Fundamentals of Statistics: A Brief Insight, Elite Publishers, Delhi.
2. Fisher R.A. and Yates F. (1938). Statistical Tables for Biological, Agricultural, and Medical Research, Oliver and Boyd, Edinburgh.
3. Goon A.M., Gupta M.K. and Dasgupta B. (2001). Fundamentals of Statistics (Vol. 2), Word Press.
4. Hague P.N. (1993). Questionnaire Design. London: Kogan Page.
5. Keyfitz N. and Beckman J.A. (1927). Demography Through Problems S-Verlag New York.
6. O'Hagan A. (1994). Kendall's Advanced Theory of Statistics: Volume 2B, Bayesian Inference. Edward Arnold, London.
7. Oppenheim A.N. (1997). Questionnaire Design and Attitude Measurement, 2nd Edition, London: Heinemann.
8. Pearson K. (1914). Tables for Statisticians and Biometricians, Cambridge University Press.
9. Rowntree D. (1981). Statistics Without Tears. An Introduction for Non-mathematicians, Penguin Books, London.
10. Tarray T.A. (2016). Statistical Sample Survey Methods and Theory, Elite Publishers, Delhi.
11. Gupta, S.C. and Kapoor, V.K. (2014). Fundaments of Mathematical Statistics, Sultan Chand and Sons.
12. Singh, D. and Chaudhary, F.S. (2004). Theory and Analysis of Sample Survey Design, New Age International Private Limited.

❑❑❑

Chapter 6

Economics of Production of Capsicum (*Capsicum annum*) in Solan District of Himachal Pradesh

Subhash Sharma*, Sangya Kumari, Amit Guleria, RS Prasher, MK Vaidya and Ravinder Sharma

Deptt. of Social Sciences, Dr YS Parmar University of Horticulture and Forestry, Nauni, Solan-173230, HP, Email: dsubhashsharma@gmail.com

ABSTRACT

The study estimated the Economics of Production of Capsicum (*Capsicum annum*) in Solan District of Himachal Pradesh was carried out in Kandaghat and Solan blocks by taking representative sample of 80 farmers having 40 farmers from each block. The results revealed that the area production and productivity of capsicum in H.P and Solan has shown significant increase over the years. The cost of cultivation of capsicum on per farm basis varied from Rs. 19575 in marginal farm category to Rs. 29196 in medium farm category. The per hectare cost of cultivation of capsicum was Rs. 134896 at an overall level and varied from Rs. 130987 in semi-medium farm category to Rs. 137509 in marginal farm category. Cost of capsicum production at an overall level was Rs. 776.81 per quintal. Per hectare gross returns varied from Rs. 289178.13 in medium farm category to Rs. 354118.83 in marginal farm category. The input-output ratio at an overall level was found to be 2.39, which mean that at an investment of Rs. 1 on capsicum cultivation the farmers in the study area were getting Rs. 2.39 as returns at an overall level.

Keywords: Capsicum, production, costs, returns, Himachal Pradesh.

Introduction

Capsicum (*Capsicum annum*) is called as 'sweet pepper' or 'shimla mirch' and is one of the most popular and highly remunerative annual herbaceous vegetable crops. India contributes one fourth of world's production of capsicum. It is one of the cash crops in India and is grown in an area about 24 thousand hectare with an annual production of 306 thousand tonnes (NHB, 2016). In India, it is commercially grown in Himachal Pradesh, Jammu & Kashmir, Uttarakhand, Arunachal Pradesh and Darjeeling district of West Bengal during summer months and as an autumn crop in Maharashtra, Karnataka, Tamil Nadu and Bihar. Himachal Pradesh ranks first in capsicum production in India with a production of 57.41 thousand tonnes (19.90 %) followed by Karnataka with 56.70 thousand tonnes (19.66 %), Madhya Pradesh with 33.84 thousand tonnes (11.73 %) (NHB, 2016).

It is an important off-season vegetable of western Himalayas and offers potential for boosting economy of farmers of hilly regions. In Himachal Pradesh, it is extensively grown as cash crop (April-October) in agro climatic zones-I, II and III in an open environment. Top five districts in area and production of capsicum in Himachal Pradesh are Solan, Sirmaur, Kangra, Mandi and Shimla. Solan district of Himachal Pradesh ranks first in capsicum production with a total production of 34850 tonnes in 1217 hectare, followed by Sirmaur, Kangra, Mandi and Shimla. Solan and Kandaghat developmental blocks of Solan district contribute most of the capsicum production in state. The major varieties of capsicum grown in Solan are California Wonder, Yolo Wonder, Solan Bharpur, Dollar and Bharat. Capsicum consumption in India is increasing now-a-days due to increasing demand by urban consumers. There is a good demand for export too.

At present, capsicum growers in hilly regions are facing several challenges in the open field cultivation like vagaries of weather viz., fluctuating temperature, unprecedented rains and frequent hail storms which affect the yield and quality of the produce, thereby reducing the profit margin of producers. Due to these challenges, the scenario of capsicum production has been changing significantly. The protected cultivation is getting preference over open field cultivation for off-season quality production by extending availability of the quality produce, higher productivity and improved nutritional attributes of the polyhouse produce (Sreedhara, 2010). There are not much studies conducted on economic aspects of capsicum production in open conditions in HP. Not much economic analysis has been done on the production of capsicum under open conditions. There has been increased demand for capsicum by the consumers and number of farmers are also showing interest in the cultivation of this crop. The knowledge of the costs and returns structure of capsicum is crucial for policy formulation, selecting appropriate production strategies and identifying regional comparative advantage in its production. The reliable information on costs and returns of the crop helps the farmers in allocating the scarce resources most optimally. In the backdrop of this, the study was conducted to study the economics of production of Capsicum (*Capsicum annum*) in Solan district of Himachal Pradesh.

Methodology

Solan district of Himachal Pradesh was purposively selected for the present study because of its significant contribution to area and production of capsicum. Multistage Random sampling was adopted to select the ultimate sample of the respondents i.e. the capsicum growers. At the first stage, two development blocks i.e. Kandaghat and Solan, out of 5 blocks were selected. At the second stage, a list of villages growing capsicum in the selected blocks was prepared and 8 villages from each block were randomly selected. At the third stage, 5 capsicum growers in each of the selected villages were selected for collection of primary data. Thus, a sample of 80 growers were selected for the study. For the analysis of data, the total vegetable growers were divided into four classes according to their size of land holding, viz., marginal (<1 ha), small (1-2 ha), semi-medium (2-4 ha) and medium farmers (4-10 ha).

Costs and Returns Analysis

CACP Cost Concepts

- Cost A_1

Cost A_1 includes:

1. Cost of hired human labour
2. Cost of bullock labour
3. Cost of owned machinery
4. Cost of hired machinery
5. Cost of fertilizers
6. Cost of manures

7. Cost of seed (owned/purchased)
8. Cost of plant protection chemicals
9. Land revenue
10. Depreciation of farm machinery, equipments and farm buildings
11. Interest on owned working capital:
    - Cost A_2 = Cost A_1 + Rent paid for leased in land
    - Cost B_1 =Cost A_1 + Interest on owned fixed capital assets excluding land
    - Cost B_2 = Cost B_1 + Rental value of own land (less land revenue) + Rent paid for leased in land
    - Cost C_1 = Cost B_1 + Imputed value of family labour
    - Cost C_2= Cost B_2 + Imputed value of family labour
    - Cost C_3= Cost C_2 +10 per cent of cost C_2 on account of managerial function performed by the farmer.

Income measures

For working out profitability of capsicum in the study area, following income measures were worked out:

a) Farm Investment Income (FII)

   FII = Farm Business Income – Imputed value of family labour

b) Family Labour Income (FLI)

   It is the return to family labour (including management).

   F.L.I. = Gross income – Cost B_2

c) Net Farm Income (NFI)

   It is the net profit after deducting all cost items *i.e.*, variable and fixed costs from gross income.

   NI = Gross income – Total cost (Cost C_3)

d) Farm Business Income (FBI)

   It is the disposable income out of the enterprise.

   FBI = Gross income – Cost A_1

Results and Discussion

Cost of Production of Capsicum

The per farm cost of capsicum production was estimated and results have been presented in Table 1. Human labour contributed the highest share (21.96 %) to the total cost of production and varied from Rs. 4242.54 (marginal farm category) to Rs. 6110.43 (medium farm category). The Bullock labour varied from Rs. 93 to Rs. 403 from marginal to medium farm category. The cost of hired machinery was found highest for semi-medium farm category (Rs. 500) followed by marginal (Rs. 458.82), medium (Rs. 366.66) and small farm category (Rs. 257.14). The cost of planting material varied from Rs. 935.59 in marginal farm category to Rs. 1815 in medium farm category. Fertilizers cost was highest in medium farm category i.e. Rs. 1339.75 and lowest in marginal farm category i.e. Rs. 714.66. The cost for plant protection chemicals ranged from Rs. 1018.09 in marginal farm category to Rs. 1370.83 in medium farm category. Staking material cost varied from Rs. 808.82 in marginal farm category to Rs. 1583.33 in medium farm category. The cost for FYM was found to be the highest in marginal farm category i.e. Rs. 4012.50 and lowest in medium farm category i.e. Rs. 3725. Among different input on overall level, the per farm cost was observed highest for human labour (Rs. 4850.54) followed by FYM (Rs. 3760.50), planting material (Rs. 1181.50), plant protection chemicals (Rs. 1176.44), staking material (Rs. 984.38), fertilizer (Rs. 848.36), hired machinery (Rs.387.50) and bullock labour (Rs. 151.13). The total working capital contributed 45.28 per cent to the total cost at an overall level and varied from 43.33 per cent in marginal farm category to 52.25 per cent in medium farm category.

The total variable cost varied from 53.26 per cent in marginal farm category to 64.22 per cent in medium farm category and was Rs. 10425.94 in marginal farm, Rs. 12811.95 in small, Rs. 13637.90 in semi-medium and Rs. 18750.79 in medium farm category. Total fixed cost varied from 27.32 per cent in marginal farm category to 31.55 per cent in semi-medium farm category and was Rs. 5346.97, Rs. 6683.45, Rs. 7714.38 and Rs. 8986.56 for marginal, small, semi-medium and medium farm category, respectively.

Table 1: Farm category wise cost of capsicum cultivation in sampled households

(Rupees per farm)

Item of cost	Marginal	Small	Semi-medium	Medium	Overall
1.Variable cost					
a) Human labour	4242.54 (21.67)	5062.03 (22.39)	5449.78 (22.29)	6110.43 (20.93)	4850.54 (21.96)
i) Family labour	3801.91 (19.42)	3111.01 (13.76)	3098.59 (12.67)	1459.13 (5.00)	3278.89 (15.35)
ii) Hired labour	440.64 (2.25)	1951.02 (8.63)	2351.20 (9.62)	4651.30 (15.93)	1571.66 (6.61)
b) Bullock labour	93.00 (0.48)	205.93 (0.91)	62.00 (0.25)	403.00 (1.38)	151.13 (0.66)
c) Tiller/tractor (hired)	458.82 (2.34)	257.14 (1.14)	500.00 (2.04)	366.66 (1.26)	387.50 (1.80)
d) Material Cost					
i) Planting material	935.59 (4.78)	1325.00 (5.86)	1226.67 (5.02)	1815.00 (6.22)	1181.50 (5.30)
ii) Fertilizers	714.66 (3.65)	837.58 (3.71)	1006.63 (4.12)	1339.75 (4.59)	848.36 (3.81)
iii) Plant protection chemicals	1018.09 (5.20)	1272.86 (5.63)	1302.92 (5.33)	1370.83 (4.70)	1176.44 (5.33)
iv) Staking material	808.82 (4.13)	1080.36 (4.78)	958.33 (3.92)	1583.33 (5.42)	984.38 (4.42)
v) FYM	4012.50 (20.50)	3493.39 (15.45)	3687.50 (15.08)	3725.00 (12.76)	3760.50 (17.34)
vi) Total material cost	7489.66 (38.26)	8009.19 (35.43)	8182.04 (33.46)	9833.92 (33.68)	7951.17 (36.21)
e) Total working capital	8482.12 (43.33)	10423.28 (46.11)	11095.24 (45.38)	15254.88 (52.25)	10061.45 (45.28)
f) Interest on working capital	247.40 (1.26)	304.01 (1.34)	323.61 (1.32)	444.93 (1.52)	293.46 (1.32)
g) Risk margin	848.21 (4.33)	1042.33 (4.61)	1109.52 (4.54)	1525.49 (5.22)	1006.15 (4.53)
h) Managerial cost	848.21 (4.33)	1042.33 (4.61)	1109.52 (4.54)	1525.49 (5.22)	1006.15 (4.53)
Total Variable cost	10425.94 (53.26)	12811.95 (56.67)	13637.90 (55.78)	18750.79 (64.22)	12367.20 (55.66)
2. Fixed cost					
a) Land revenue	4.45 (0.02)	5.31 (0.02)	5.83 (0.02)	6.67 (0.02)	5.13 (0.02)
b) Depreciation	1062.64 (5.43)	1476.10 (6.53)	1859.41 (7.60)	2259.31 (7.74)	1416.62 (6.31)

c) Interest on fixed capital	481.96 (2.46)	666.49 (2.95)	868.92 (3.55)	1028.91 (3.52)	645.61 (2.88)
d) Rental value of land	3797.93 (19.40)	4535.55 (20.06)	4980.21 (20.37)	5691.67 (19.49)	4375.47 (19.79)
Total fixed cost	5346.97 (27.32)	6683.45 (29.56)	7714.38 (31.55)	8986.56 (30.78)	6442.82 (29.00)
Total cost	19574.82 (100.00)	22606.41 (100.00)	24450.86 (100.00)	29196.48 (100.00)	22088.91 (100.00)

Figures in parentheses are percentage to total

The per hectare cost of capsicum cultivation was estimated and results have been presented in Table 2. It is clear from the table that the cost of capsicum cultivation was the highest in marginal farm category (Rs. 137509.09) followed by medium (Rs. 136858.50), small (Rs. 132978.89) and semi-medium farm category (Rs. 130986.75). Human labour contributed the highest share (21.96 %) to the total cost of cultivation and varied from Rs. 28642.66 in medium farm category to 29803 in marginal farm category. The contribution of bullock labour to the total cost varied from 0.48 per cent in marginal farm category to 1.38 per cent in medium farm category. The cost of hired machinery was found the highest for marginal farm category (Rs. 3223.14) followed by semi-medium (Rs. 2678.57), medium (Rs. 1718.72) and small farm category (Rs. 1512.61) respectively. The cost of planting material varied from Rs. 6571.43 in semi-medium farm category to Rs. 8507.81 in medium farm category. Fertilizer cost was the highest for medium farm category i.e. Rs. 6280.08 and lowest for small farm category i.e. Rs. 4926.94. The cost of plant protection chemicals ranged from Rs. 6425.78 in medium farm category to Rs. 7487.39 in small farm category. Staking material cost varied from 3.92 per cent in semi-medium farm category to 5.42 per cent in medium farm category. FYM was contributing 17.34 per cent to the total cost at an overall level. Among different inputs on overall level, the per farm cost was observed highest for human labour (Rs. 29615.60) followed by FYM (Rs. 23444.49), plant protection chemicals (Rs. 7189.05), planting material (Rs. 7144.97), staking material (Rs. 5965.77), fertilizers (Rs. 5137.98), hired machinery (Rs. 2429.94) and bullock labour (Rs. 1309.37), respectively. The total working capital contributed 45.28 per cent to the total cost at an overall level and varied from 43.33 per cent in marginal farm category to 52.25 per cent in medium farm category.

The contribution of total variable cost to the total cost was 53.26 per cent in marginal farm category, 56.67 per cent in small, 55.78 per cent in semi-medium and 64.22 per cent in medium farm category. The contribution of fixed cost to the total cost was 27.32 per cent in marginal farm category, 29.56 per cent for small, 31.55 per cent for semi-medium and 30.78 per cent for medium farm category.

Table 2: Farm category-wise cost of capsicum cultivation in sampled households

(Rupees per hectare)

Item of cost	Marginal	Small	Semi-medium	Medium	Overall
1.Variable cost					
a) Human labour	29803.00 (21.67)	29776.67 (22.39)	29195.26 (22.29)	28642.66 (20.93)	29615.60 (21.96)
i) Family labour	26707.60 (19.42)	18300.05 (13.76)	16599.57 (12.67)	6839.67 (5.00)	20758.66 (15.35)
ii) Hired labour	3095.40 (2.25)	11476.62 (8.63)	12595.69 (9.62)	21802.99 (15.93)	8856.94 (6.61)
b) Bullock labour	653.31 (0.48)	1211.34 (0.91)	332.14 (0.25)	1889.06 (1.38)	893.13 (0.66)
c) Tiller/tractor (hired)	3223.14 (2.34)	1512.61 (1.14)	2678.57 (2.04)	1718.72 (1.26)	2429.94 (1.80)
d) Material Cost					
i) Planting material	6572.31 (4.78)	7794.12 (5.86)	6571.43 (5.02)	8507.81 (6.22)	7144.97 (5.30)
ii) Fertilizers	5020.35 (3.65)	4926.94 (3.71)	5392.63 (4.12)	6280.08 (4.59)	5137.98 (3.81)
iii) Plant protection chemicals	7151.86 (5.20)	7487.39 (5.63)	6979.91 (5.33)	6425.78 (4.70)	7189.05 (5.33)
iv) Staking material	5681.82 (4.13)	6355.04 (4.78)	5133.93 (3.92)	7421.88 (5.42)	5965.77 (4.42)
v) FYM	28186.98 (20.50)	20549.37 (15.45)	19754.46 (15.08)	17460.94 (12.76)	23444.49 (17.34)
vi) Total material cost	52613.33 (38.26)	47112.87 (35.43)	43832.37 (33.46)	46096.48 (33.68)	48882.26 (36.21)
e) Total working capital	59585.17 (43.33)	61313.43 (46.11)	59438.77 (45.38)	71507.25 (52.25)	61062.26 (45.28)
Item of cost	**Marginal**	**Small**	**Semi-medium**	**Medium**	**Overall**
f) Interest on working capital	1737.90 (1.76)	1788.31 (1.75)	1733.63 (1.69)	2085.63 (1.80)	1780.98 (1.75)
g) Risk margin	5958.52 (4.33)	6131.34 (4.61)	5943.88 (4.54)	7150.73 (5.22)	6106.23 (4.53)
h) Managerial cost	5958.52 (4.33)	6131.34 (4.61)	5943.88 (4.54)	7150.73 (5.22)	6106.23 (4.53)
Total Variable cost	73240.11 (53.26)	75364.43 (56.67)	73060.16 (55.78)	87894.33 (64.22)	75055.70 (55.66)

2. Fixed cost					
a) Land revenue	31.25 (0.02)	31.25 (0.02)	31.25 (0.02)	31.25 (0.02)	31.25 (0.02)
b) Depreciation	7464.81 (5.43)	8682.97 (6.53)	9961.15 (7.60)	10590.51 (7.74)	8500.04 (6.31)
c) Interest on fixed capital	3385.63 (2.46)	3920.50 (2.95)	4654.93 (3.55)	4823.04 (3.52)	3871.04 (2.88)
d) Rental value of land	26679.69 (19.40)	26679.69 (20.06)	26679.69 (20.37)	26679.69 (19.49)	26679.69 (19.79)
Total fixed cost	37561.38 (27.32)	39314.41 (29.56)	41327.02 (31.55)	42124.49 (30.78)	39082.02 (29.00)
Total cost	137509.09 (100.00)	132978.89 (100.00)	130986.75 (100.00)	136858.50 (100.00)	134896.38 (100.00)

Figures in parentheses are percentage to total

Cost Concepts in Capsicum Cultivation

The cost of capsicum cultivation under different cost concepts was analyzed and presented in Table 3. Table revealed that highest per farm cost (cost C_3) was observed on medium farm category (Rs. 29196.48) followed by semi-medium (Rs. 24450.86), small (Rs. 22606.41) and marginal (Rs.19574.82) where at overall level, it was Rs. 22088.91 per hectare. Cost A_1, A_2, B_1, B_2, C_1, and C_2 at an overall level were Rs.11776.66, Rs.11776.66, Rs. 12422.26, Rs. 16797.73, Rs. 15701.15, and Rs. 20076.62 respectively. Cost A_1 for medium farm category was found the highest i.e. Rs. 17965.79 followed by semi-medium (Rs. 13284.79), small (Rs. 12208.71) and marginal farm category (Rs. 9796.61). Same pattern was followed by cost A_2, B_1, B_2, C_1, C_2 and C_3 respectively.

Per hectare cost concepts in capsicum cultivation are presented in Table 4. It was observed from the table 4.16 that the highest cost (Cost C_3) was observed on marginal farm category (Rs. 137509.09) followed by medium (Rs. 136858.50), small (Rs. 132978.89) and semi-medium farm category (Rs. 130986.75) where at an overall level it was Rs. 134896.38 per hectare. Cost A_1, A_2, B_1, B_2, C_1, and C_2 at an overall level were found to be Rs.71374.54, Rs. 71374.54, Rs. 75245.58, Rs. 10192558, Rs. 96004.24 and Rs. 122683.92 respectively.

Table 3: Farm category-wise cost concepts in capsicum cultivation

(Rupees/farm)

Size of Farm	Cost Concepts						
	A_1	A_2	B_1	B_2	C_1	C_2	C_3
Marginal	9796.61	9796.61	10278.56	14076.49	14080.47	17878.40	19574.82
Small	12208.71	12208.71	12875.20	17410.75	15986.21	20521.75	22606.41
Semi-medium	13284.10	13284.10	14153.02	19133.22	17251.60	22231.81	24450.86
Medium	17965.79	17965.79	18994.71	24686.37	20453.84	26145.50	29196.48
Overall	11776.66	11776.66	12422.26	16797.73	15701.15	20076.62	22088.91

Table 4: Farm category-wise cost concepts in capsicum cultivation

(Rupees/hectare)

Size of Farm	Cost Concepts						
	A_1	A_2	B_1	B_2	C_1	C_2	C_3
Marginal	68819.14	68819.14	72204.77	98884.46	98912.37	125592.06	137509.09
Small	71815.96	71815.96	75736.46	102416.15	94036.52	120716.21	132978.89
Semi-medium	71164.80	71164.80	75819.73	102499.42	92419.30	119098.99	130986.75
Medium	84214.64	84214.64	89037.68	115717.37	95877.36	122557.04	136858.50
Overall	71374.54	71374.54	75245.58	101925.26	96004.24	122683.92	134896.38

Net returns over cost concepts in capsicum

Net returns are the returns to the producer after all production expenses have been paid. The net returns over different cost concepts were analyzed and presented in Table 5. Per farm net returns over different cost concepts in capsicum were analyzed and presented in Table 6. Net returns over cost C_3 were the highest on medium farms (Rs. 32032.35), followed by semi-medium (Rs. 29647.77), small (Rs. 28985.90) and marginal farms (6208.19). At an overall level, net returns for cost C_3 were Rs. 19633.14.

The per hectare net returns over cost concepts in capsicum are presented in Table 4.18. The net returns for cost C_3 were highest on marginal farm category (Rs. 216609.74) followed by small farms (Rs. 173273.84), semi-medium (Rs. 160561.69), medium (Rs. 152319.63). At an overall level the net returns for C_3 were found to be Rs. 188213.21. For cost A_1, A_2, B_1, B_2, C_1, C_2 and C_3 the net returns were highest in case of marginal farm category i.e. Rs. 285299.69, Rs. 285299.69, Rs. 281914.06, Rs. 255234.37, Rs. 255206.46, Rs. 228526.77 and Rs. 216609.74, respectively.

Table 5: Farm category-wise net returns over cost concepts in capsicum

(Rupees/farm)

Size of Farm	Net Returns						
	A_1	A_2	B_1	B_2	C_1	C_2	C_3
Marginal	15986.41	15986.41	15504.45	11706.52	11702.55	7904.62	6208.19
Small	39383.60	39383.60	38717.11	34181.57	35606.11	31070.56	28985.90
Semi-medium	40814.53	40814.53	39945.61	34965.40	36847.02	31866.81	29647.77
Medium	43263.04	43263.04	42234.13	36542.46	40775.00	35083.33	32032.35
Overall	29945.39	29945.39	29299.78	24924.31	26020.90	21645.43	19633.14

Table 6: Farm category-wise net returns over cost concepts in capsicum

(Rs./hectare)

Size of farm	Net returns						
	A_1	A_2	B_1	B_2	C_1	C_2	C_3
Marginal	285299.69	285299.69	281914.06	255234.37	255206.46	228526.77	216609.74
Small	234436.77	234436.77	230516.27	203836.58	212216.21	185536.53	173273.84
Semi-medium	220383.64	220383.64	215728.71	189049.02	199129.13	172449.45	160561.69
Medium	204963.48	204963.48	200140.44	173460.75	193300.77	166621.08	152319.63
Overall	251735.05	251735.05	247864.01	221184.32	227105.35	200425.66	188213.21

Cost of Capsicum Production

Cost of capsicum production was analyzed and presented in Table 7. The per hectare cost of cultivation was Rs. 137509.09, Rs. 132978.89, Rs. 130986.75 and Rs. 136858.50 for marginal, small, semi-medium and medium farms, respectively. Yield of capsicum was found to be 191.42, 165.54, 157.59 and 156.31 quintals for marginal, small, semi-medium and medium farms, respectively. The cost of production per quintal for marginal, small, semi-medium and medium farm categories was found to be Rs. 718.38, Rs. 803.29, Rs. 831.17 and Rs. 875.54, respectively. At an overall level it was found to be Rs. 776.81.

Table 7: Farm category-wise cost of capsicum production

Particulars	Marginal	Small	Semi-medium	Medium	Overall
Cost of cultivation(Rs./ha)	137509.09	132978.89	130986.75	136858.50	134896.38
Yield (qtl/ha)	191.42	165.54	157.59	156.31	174.65
Cost of production (Rs./qtl)	718.38	803.29	831.17	875.54	776.81

Income measures

Farm category-wise different income measures are presented in the Table 8. It could be observed from the table that gross returns at an overall level and for marginal, small, semi-medium and medium farm categories was Rs. 323109.58, Rs. 354118.83, Rs. 306252.73, Rs. 291548.44 and Rs. 289178.13, respectively. Net farm income in case of marginal, small, semi-medium and medium farm categories was Rs. 216685.93, Rs. 175485.56, Rs. 161796.35 and Rs. 150735.16, respectively. The farm business income was the highest for marginal farm category (Rs. 285299.69) followed by small (Rs. 234436.77), semi-medium (Rs. 220383.64) and medium farm category (Rs. 204963.48). Family labour income for marginal, small, semi-medium and medium farm categories was Rs. 255234.37, Rs. 203836.58, Rs. 189049.02, Rs. 173460.75, respectively. Farm investment income for marginal, small, semi-medium and medium farm categories was Rs. 258592.09, Rs. 216136.72, Rs. 203784.06 and Rs. 198123.81 respectively. The output input ratio in case of overall farms category was found to be 2.39 and varied from 2.11 in medium farm category to 2.58 in marginal farm category which means that at an investment of Rs. 1 on capsicum cultivation, the farmers in the study area were getting Rs. 2.39 as returns at an overall level.

Table 8: Farm category-wise income measures of capsicum in sampled households

(Rupees/ha)

Particular	Marginal	Small	Semi-medium	Medium	Overall
Gross return	354118.83	306252.73	291548.44	289178.13	323109.58
Net farm income	216609.74	173273.84	160561.69	152319.63	188213.21
Farm business income	285299.69	234436.77	220383.64	204963.48	251735.05
Family labour income	255234.37	203836.58	189049.02	173460.75	221184.32
Farm investment income	258592.09	216136.72	203784.06	198123.81	230976.39
Output input ratio	2.58	2.30	2.23	2.11	2.39

Summary

The per farm cost of capsicum cultivation at an overall level was found to be Rs. 22088.91. The human labour contributed maximum (21.96 %) to the total cost. The total material cost formed 36.21 per cent of the total cost. The cost of capsicum cultivation for marginal, small, semi-medium and medium farmers was estimated to be Rs. 19575, Rs. 22606, Rs. 24451 and Rs. 29196, respectively. Cost of capsicum cultivation was estimated to be the highest for medium farm category followed by semi-medium, small and marginal farm category. The per hectare cost of capsicum cultivation at an overall level was found to be Rs. 134896. The cost of capsicum cultivation for marginal, small, semi-medium and medium farmers was estimated to be Rs.

137509, Rs. 132979, Rs. 130987 and Rs. 136859, respectively. The per hectare cost of capsicum production was estimated to be highest for marginal farm category followed by medium, small and semi-medium farm category. At an overall level, cost A_1, A_2, B_1, B_2, C_1, C_2 and C_3 were Rs. 11777, Rs.11777, Rs. 12422, Rs. 16798, Rs. 15701, Rs. 20077 and Rs. 22089 per farm, respectively. At an overall level, per hectare cost of capsicum cultivation (C_3) was found to be Rs. 134896.38, out of which cost A_1, A_2, B_1, B_2, C_1, and C_2 were found to be Rs. 71375, Rs. 71375, Rs. 75246, Rs. 101925, Rs. 96004 and Rs. 122684, respectively. Per farm net returns for cost A_1, A_2, B_1, B_2, C_1, C_2 and C_3 were Rs. 29945, Rs. 29945, Rs. 29300, Rs. 24924, Rs. 26021, Rs. 21645 and Rs. 19633, respectively. Per hectare net returns for costs A_1, A_2, B_1, B_2, C_1, C_2 and C_3 were Rs. 251735, Rs. 251735, Rs. 247864, Rs. 221184, Rs. 227105, Rs. 200426 and Rs. 188213, respectively. The per hectare production of capsicum varied from 156.31 quintals in medium farm category to 191.42 quintals in marginal farm category. The cost of production per quintal for marginal, small, semi-medium and medium farm categories was found to be Rs. 718, Rs. 803, Rs. 831 and Rs. 876, respectively. At an overall level, it was found to be Rs. 776. The gross returns at an overall level for capsicum were found to be Rs. 323109.58. The net farm income at an overall level was Rs. 188213.21. The input output ratio was found to be the highest (2.58) in case of marginal farmers. At an overall level input-output ratio was found to be 2.39, which mean that at an investment of Rs. 1 on capsicum cultivation the farmers in the study area were getting Rs. 2.39 as returns at an overall level.

References

1. National Horticulture Board. 2016. *Horticultural Data Base,* 2016. National Horticulture Board, New Delhi.
2. Sreedhara DS. 2010. Comparative economics of capsicum production under protected and open conditions in northern Karnataka. M.Sc. Thesis. Department of Agricultural Economics, University of Agricultural Sciences, Dharwad, pp. 1-72.

❑❑❑

Chapter 7

Economics of Production of Mango: A Case Study of Kangra District of Himachal Pradesh

Subhash Sharma*, Parul Sharma, Amit Guleria, RS Prasher, MK Vaidya and Ravinder Sharma

Dept. of Social Sciences, Dr YS Parmar University of Horticulture and Forestry, Nauni, Solan-173230, HP, Email: dsubhashsharma@gmail.com

ABSTRACT

The study estimated the economics of production of mango in Kangra district of Himachal Pradesh. A sample of 80 farmers cultivating mango was selected using multistage simple random sampling. The farmers were categories on the basis of their total land holding as marginal (<1 ha), small (1-2) and medium (2-10) and as overall. Total cost of mango cultivation per hundred plants was estimated to Rs.34940.45 at overall level and practically found almost same in all categories of farmers. The maintenance cost of non-bearing plants was Rs. 38391.13, Rs.40998.07, Rs. 44050.47, Rs. 47271.17 and Rs. 50540.90 per hundred plants in first to fifth year of age, respectively. Maintenance cost during bearing stage was increasing with age between Rs. 56873.29 and 57861.69 per hundred plants in early and main bearing stage, respectively and then gradually decreased to Rs. 55536.36 in late bearing stage. The proportion of variable cost ranged from 54.64 to 56.50 per cent in different years and share of fixed cost varied 53.50 to 45.36 per cent in the total cost.

Keywords: Mango, Production, Costs, Returns, Himachal Pradesh.

Introduction

Himachal Pradesh is having the advantage of varied climate, ranging from subtropical to dry temperate. The total fruit production in the state is 9,28,829 metric tonnes from an area of 2,26,799 hectares.

The area under mango cultivation in the state is 41,523 hectares which accounts for 18.30 per cent of total area under fruits in H.P. and production is 37,628 metric tonnes which is 4 per cent of total fruit production of H.P. (NHB, 2016). Mango cultivation is concentrated in Kangra, Una, Bilaspur, Hamirpur and parts of Mandi district of H.P. Kangra district is the major mango growing district of the state covering an area of 21,286 hectares, which is 51.26 per cent of total area under mango cultivation in H.P. and producing 24,900 metric tonnes i.e. 47.70 per cent of total production of Himachal Pradesh (Directorate of Economics and Statistics, 2016).

Production of mango is distinguished from annual crops by the long gestation period, an extended period of output flow and varying stages of productivity over the lifetime (Chand, 1994). Therefore, it is more difficult to determine economics of growing mango as compared to the annual crops. Having sufficient awareness regarding profitability of any enterprise is needed to make rational decisions at the farm level. The producers would like to know the results of this economic activity by working out detailed costs and benefits of the investment in the project (Akcay & Uzunoz, 2005). Unfortunately, the farmers know very little about economics of growing mango in the state. The farmers need information regarding investment and returns from mango cultivation.

With this background, the study was conducted to estimate the economics of production of mango in Kangra district of Himachal Pradesh.

Methodology

Kangra district of Himachal Pradesh was purposively selected because of its significant contribution in area and production of mango in Himachal Pradesh. Multistage Random Sampling technique was used to select the ultimate respondents i.e. the mango growers. At first stage, two blocks of Kangra district (Nurpur and Indora) were selected on the basis of area under mango cultivation. At the second stage, a list of the villages in the selected blocks was prepared and 8 villages from each selected blocks were selected. At the third stage, a list of farmers having mango orchards of the selected villages was prepared with the help of the officials from the Revenue and Horticulture departments and 5 mango growers were selected from each village, forming a sample of 80 farmers for the present study.

After the field survey, for the analysis of data, the total respondents were divided into three classes according to the size of their land holdings, viz., marginal (<1 ha), small (1-2 ha) and medium farmers (2-10 ha).

To meet out the requirements of the objective, tabular analysis averages, percentages, costs and returns analysis were used.

Prorated establishment cost

The establishment cost is distributed over the bearing life (n) using principle of amortization at given rate of interest and is charged at 5 per cent and is amortized for 60 years during the bearing stage, with the following formula:

$$P = E.C. \frac{i}{1-(1+i)^{-n}}$$

Where,

P = Amount of annual amortized cost

E.C. = Establishment cost (cost during non-bearing stage)

n = Number of productive life period of crop in years

i = Interest rate

Costs and returns

In determining the costs and returns of perennial crops, age of plantation plays an important role. Keeping in view of this point, the sample has been further classified into different groups according to age of mango plants. First group consisting of the year of plantation considered as zero year showing the initial cost of plantation. Second group consisted of non-bearing plants from 1-5 years considered as non-bearing stage and third group consisted of bearing stage which has been further divided into three sub groups i.e. early bearing (6-15 years), main bearing (16-35 years) and late bearing (>35 years).

The total age of mango plantation was considered to be 60 years. Following cost concepts were used:

Plantation cost: All costs incurred by mango growers from preparation of land upto planting of plants.

i) Plantation cost = Land preparation (pit digging + layout) + Cost of plants and other inputs.

ii) Fixed cost = Land revenue + Rental value of land + Depreciation + Interest on fixed capital

iii) Variable cost = Cost of inputs used including labour + interest on working capital

iv) Establishment cost = Plantation cost + cost during non-bearing stage.

v) Maintenance cost = Establishment cost and cost during bearing stage

Results and Discussion

Costs and Returns of Mango Cultivation

The cost of cultivation is a subject of attention to a wide range of users of cost data and it assumes particular importance in the area of planning. The cost of cultivation of horticultural commodities for planning is appreciated as these data inform the planners about the area, where it is economical to produce various commodities and the regions that would accordingly be most appropriate for the development of industries based on the agricultural raw material. These facilitates the experts to make practical recommendations for farm planning aimed at better allocation of existing resources and introduction of improved agronomic practices that would increase the competence in mango production. The summation of cost incurred during plantation (initial investment) and costs during non-bearing stage (establishment cost) which is distributed over total productive life of the plantation.

The costs and returns were worked out on hundred plant basis under marginal, small, medium and overall farm categories. For estimating the costs and returns from mango, it has been assumed that:

i) First bearing starts from 6th year in the study area.

ii) The major operations and inputs requirement remains same in early bearing stage (6- 15 year), main bearing stage (16-35 year), and late bearing stage (> 35 year).

iii) Total economic life of the plantation is 60 years in the study area.

The above mentioned stages are based on growth and productivity pattern of the plant. In mango, initial investment is quite heavy for reasons of the cost involved in land development, digging of pits, application of manure & fertilizers and material cost. Growers have to incur cost on maintenance for about 5th years without any returns.

Cost of plantation (Initial investment)

Cost of plantation consists of all the expenditures incurred on different inputs and labour required for the establishment of an orchard. It comprises of specifically expenditure on land preparation and digging of pits along with FYM and planting cost. It also includes cost on human labour (hired and family), land revenue, risk margin and managerial cost.

The item-wise plantation cost incurred in the initial year has been presented in Table 1. It is clear from the Table 4.18 that, mango orchardists in study area incurred on an average, total cost of Rs.34940.45 per hundred plants. The per cent share of planting material to the total cost was maximum (23.64%) followed by FYM (4.14%) and soil treatment (0.80%). Human labour contributing 21.56 per cent to the total cost at overall level, which varied from 19.84 per cent in marginal farm to 24.18 per cent in medium farm category. Per cent share of interest on working capital was 1.75 per cent and risk margin & managerial cost contributed equally to the total cost was i.e. 3.23 per cent at an overall level. Total variable cost varied from 58.24 per cent in marginal farm to 59.94 per cent in medium farm and at overall level was 59.13 per cent. Total fixed cost was 41.76 per cent on marginal farm, 41.01 per cent on small farm and 40.06 per cent on medium farm category, respectively. Rental value of land shared 33.70 per cent of the total cost at overall level. Total cost varied from Rs. 32168.29 to Rs. 37820.39 among different farm categories.

Table 1: Initial costs of mango orchard in the study area

(Rupees/100 Plants)

Sr. No.	Particulars	Initial investment			
		Marginal	Small	Medium	Overall
1	**Material Cost**				
a)	Planting material	8235.29 (25.60)	8250.00 (23.88)	8312.50 (21.98)	8259.97 (23.64)
b)	FYM	1372.55 (4.27)	1416.67 (4.10)	1600.00 (4.23)	1445.49 (4.14)
c)	Soil treatment	274.51 (0.85)	279.17 (0.81)	287.50 (0.76)	279.32 (0.80)
	Total material cost	9882.35 (30.72)	9945.83 (28.79)	10200.00 (26.97)	9984.78 (28.58)
2	**Human Labour**				
a)	Family labour	6382.35 (19.84)	7218.75 (20.89)	6037.50 (15.96)	6542.75 (18.73)
b)	Hired labour	0.00 (0.00)	583.33 (1.69)	3106.25 (8.21)	990.39 (2.83)
	Total human labour	6382.35 (19.84)	7802.08 (22.58)	9143.75 (24.18)	7533.14 (21.56)
3	Interest on working capital	494.12 (1.54)	526.46 (1.52)	665.31 (1.76)	548.76 (1.57)
4	Risk Margin	988.24 (3.07)	1052.92 (3.05)	1330.63 (3.52)	1129.63 (3.23)
5	Managerial cost	988.24 (3.07)	1052.92 (3.05)	1330.63 (3.52)	1129.63 (3.23)
6	Total Variable Cost	18735.29 (58.24)	20380.21 (58.99)	22670.31 (59.94)	20660.86 (59.13)
7	**Fixed Cost**				
a)	Land revenue	31.00 (0.10)	31.00 (0.09)	31.00 (0.08)	31.00 (0.09)
b)	Depreciation	1180.31 (3.67)	1719.48 (4.98)	2399.15 (6.34)	1662.01 (4.76)
c)	Interest on fixed capital	446.68 (1.39)	644.40 (1.87)	944.93 (2.50)	636.79 (1.82)
d)	Rental value of land	11775.00 (36.60)	11775.00 (34.08)	11775.00 (31.13)	11775.00 (33.70)
	Total fixed cost	13432.99 (41.76)	14169.88 (41.01)	15150.07 (40.06)	14279.60 (40.87)
8	Total cost	32168.29 (100.00)	34550.09 (100.00)	37820.39 (100.00)	34940.45 (100.00)

Figure in parentheses are percentage to total cost

Cost during non- bearing stage of mango plantation in different farm categories

The item-wise maintenance cost during non-bearing stage of mango for marginal, small, medium and overall farms were presented in Tables 2 to Table 5. The total maintenance cost of non-bearing mango plantation has shown positive relationship with age of the plants and was estimated to Rs. 38938.04, Rs. 40965.02, Rs. 43982.91 Rs. 46939.33 and Rs. 50363.15 per hundred plants in the first year to fifth year, respectively (Table 2). The per cent share of total material cost varied from 4.41 to 4.72 per cent from first to fifth year. Total labour cost was 42.24 per cent in the fifth years to 54.76 per cent in the first year. Marginal farmers used all the family labour for different operations. Total variable cost ranged between 48.14 to 60.34 per cent from fifth to first year and fixed cost varied from 39.66 per cent to 51.86 per cent from first to fifth year. Risk margin and managerial cost exhibited positive trends with the age of plantation. The per cent share of depreciation cost and rental value of land in the total cost declined with the increase in the age of plants; however, absolute value remains the same.

Material cost contributed 4.91 per cent to 6.46 per cent of the total cost in the fifth year to first year, respectively (Table.3). Total labour cost varied from 44.13 per cent to 46.70 per cent in the first to fifth year. Per cent share of interest on working capital was 1.75 per cent and risk margin and managerial cost contributed equally to the total cost was i.e 0.88 per cent at an overall level. While interest on past establishment cost was found to have positive relationship with the age during non-bearing years. Renal value of land was 30.79 per cent, 28.62 per cent, 26.70 per cent, 24.68 per cent and 23.23 per cent from first to fifth year, respectively. Rental value of land in the total cost declined with the increase age of plants. Total cost was estimated to Rs. 38247.54, Rs. 41145.70, Rs. 44098.16, Rs. 47716.19 and Rs. 50688.59 per hundred plants in the first year to fifth year, respectively (Table 4).

In case of medium category, total cost was Rs.37643.71, Rs. 40884.44, Rs. 44108.56, Rs. 47315.67 and Rs. 50668.37 per hundred plants from the first to fifth year, respectively (Table 4). The per cent share of FYM in total cost was maximum followed by plant protection chemicals and Bordeaux mixture. The total material cost varied from 5.21 per cent to 6.59 per cent from fifth to first year .Total labour cost in medium farm varied from 34.81 per cent in the fifth year to 41.69

per cent in first year. As expected interest on working capital exhibited positive relation with age while interest on fixed capital decreased with age of plantation.

Table 2: Maintenance cost during non- bearing stage of mango orchard in case of marginal farms

(Rupees/100 Plants)

Sr. No.	Particulars	Age of plantation in years				
		1st year	2nd year	3rd year	4th year	5th year
1	**Material Cost**					
a)	Gap filling	83.33 (0.21)	-	-	-	-
b)	Bordeaux mixture	125.00 (0.32)	181.82 (0.44)	212.12 (0.48)	235.29 (0.50)	555.56 (1.10)
c)	FYM	1300.00 (3.34)	1333.33 (3.25)	1454.55 (3.31)	1529.41 (3.26)	1500.00 (2.98)
d)	Plant protection chemical	230.63 (0.59)	290.91 (0.71)	301.82 (0.69)	305.88 (0.65)	321.11 (0.64)
	Total material cost	1738.96 (4.47)	1806.06 (4.41)	1968.48 (4.48)	2070.59 (4.41)	2376.67 (4.72)
2	**Human Labour**					
a)	Family labour	21320.83 (54.76)	20830.30 (50.85)	21084.85 (47.94)	21164.71 (45.09)	21272.22 (42.24)
	Total human labour	21320.83 (54.76)	20830.30 (50.85)	21084.85 (47.94)	21164.71 (45.09)	21272.22 (42.24)
3	Interest on working capital	86.95 (0.22)	90.30 (0.22)	98.42 (0.22)	103.53 (0.22)	118.83 (0.24)
4	Risk Margin	173.90 (0.45)	180.61 (0.44)	196.85 (0.45)	207.06 (0.44)	237.67 (0.47)
5	Managerial cost	173.90 (0.45)	180.61 (0.44)	196.85 (0.45)	207.06 (0.44)	237.67 (0.47)
6	Total Variable Cost	23494.53 (60.34)	23087.88 (56.36)	23545.45 (53.53)	23752.94 (50.60)	24243.06 (48.14)
7	**Fixed Cost**					
a)	Land revenue	31.00 (0.08)	31.00 (0.08)	31.00 (0.07)	31.00 (0.07)	31.00 (0.06)
b)	Depreciation	1180.31 (3.03)	1180.31 (2.88)	1180.31 (2.68)	1180.31 (2.51)	1180.31 (2.34)
c)	Interest on fixed capital	446.68 (1.15)	446.68 (1.09)	446.68 (1.02)	446.68 (0.95)	446.68 (0.89)
d)	Rental value of land	11775.00 (30.24)	11775.00 (28.74)	11775.00 (26.77)	11775.00 (25.09)	11775.00 (23.38)
e)	Interest on past capital	2010.52 (5.16)	4444.15 (10.85)	7004.46 (15.93)	9753.39 (20.78)	12687.10 (25.19)
	Total fixed cost	15443.51 (39.66)	17877.14 (43.64)	20437.45 (46.47)	23186.38 (49.40)	26120.09 (51.86)
8	Total cost	38938.04 (100.00)	40965.02 (100.00)	43982.91 (100.00)	46939.33 (100.00)	50363.15 (100.00)

Figure in parentheses are percentage to total cost

Table 3: Maintenance cost during non- bearing stage of mango orchard in case of small farms

(Rupees/100 Plants)

Sr. No.	Particulars	Age of plantation in years				
		1st year	2nd year	3rd year	4th year	5th year
1	**Material Cost**					
a)	Gap filling	220.99 (0.58)	-	-	-	-
b)	Bordeaux mixture	234.81 (0.61)	277.78 (0.68)	291.67 (0.66)	314.21 (0.66)	316.46 (0.62)
c)	FYM	1685.08 (4.41)	1722.22 (4.19)	1755.56 (3.98)	1803.28 (3.78)	1792.41 (3.54)
d)	Plant protection chemical	331.49 (0.87)	350.22 (0.85)	361.39 (0.82)	377.05 (0.79)	379.75 (0.75)
	Total material cost	2472.38 (6.46)	2350.22 (5.71)	2408.61 (5.46)	2494.54 (5.23)	2488.61 (4.91)
2	**Human Labour**					
a)	Family labour	17377.53 (45.43)	17908.33 (43.52)	18167.59 (41.20)	18398.91 (38.56)	18111.39 (35.73)
b)	Hired labour	1160.22 (3.03)	1263.89 (3.07)	1302.78 (2.95)	1721.31 (3.61)	1949.37 (3.85)
2	Total human labour	18537.75 (48.47)	19172.22 (46.60)	19470.37 (44.15)	20120.22 (42.17)	20060.76 (39.58)
3	Interest on working capital	181.63 (0.49)	180.71 (0.44)	185.57 (0.42)	210.79 (0.44)	221.90 (0.44)
4	Risk Margin	363.26 (0.95)	361.41 (0.88)	371.14 (0.84)	421.58 (0.88)	443.80 (0.88)
5	Managerial cost	363.26 (0.95)	361.41 (0.88)	371.14 (0.84)	421.58 (0.88)	443.80 (0.88)
6	Total Variable Cost	21918.28 (57.31)	22425.97 (54.50)	22806.83 (51.72)	23668.72 (49.60)	23658.86 (46.67)
7	**Fixed Cost**					
a)	Land revenue	31.00 (0.08)	31.00 (0.08)	31.00 (0.07)	31.00 (0.06)	31.00 (0.06)
b)	Depreciation	1719.48 (4.50)	1719.48 (4.18)	1719.48 (3.90)	1719.48 (3.60)	1719.48 (3.39)
c)	Interest on fixed capital	644.40 (1.68)	644.40 (1.57)	644.40 (1.46)	644.40 (1.35)	644.40 (1.27)
d)	Rental value of land	11775.00 (30.79)	11775.00 (28.62)	11775.00 (26.70)	11775.00 (24.68)	11775.00 (23.23)
e)	Interest on past capital	2159.38 (5.65)	4549.85 (11.06)	7121.46 (16.15)	9877.59 (20.70)	12859.85 (25.37)
	Total fixed cost	16329.26 (42.69)	18719.73 (45.50)	21291.33 (48.28)	24047.47 (50.40)	27029.73 (53.33)
8	Total cost	38247.54 (100.00)	41145.70 (100.00)	44098.16 (100.00)	47716.19 (100.00)	50688.59 (100.00)

Figure in parentheses are percentage to total cost

Table 4: Maintenance cost during non- bearing stage of mango orchard in case of medium farms

(Rupees/100 Plants)

Sr.No.	Item of cost	Age of plantation in years				
		1st year	2nd year	3rd year	4th year	5th year
1	**Material Cost**					
a)	Gap filling	204.92 (0.54)	-	-	-	-
b)	Bordeaux mixture	245.90 (0.65)	273.11 (0.67)	287.23 (0.65)	366.11 (0.77)	380.00 (0.75)
c)	FYM	1557.38 (4.14)	1621.85 (3.97)	1693.62 (3.84)	1715.48 (3.63)	1720.00 (3.39)
d)	Plant protection chemical	471.52 (1.25)	495.80 (1.21)	510.64 (1.16)	535.56 (1.13)	540.00 (1.07)
	Total material cost	2479.71 (6.59)	2390.76 (5.85)	2491.49 (5.65)	2617.15 (5.53)	2640.00 (5.21)
2	**Human Labour**					
a)	Family labour	10342.21 (27.47)	10823.53 (26.47)	11125.53 (25.22)	11232.22 (23.74)	11270.00 (22.24)
b)	Hired labour	5350.41 (14.21)	5764.71 (14.10)	5957.45 (13.51)	6106.69 (12.91)	6370.00 (12.57)
	Total human labour	15692.62 (41.69)	16588.24 (40.57)	17082.98 (38.73)	17338.91 (36.65)	17640.00 (34.81)
3	Interest on working capital	391.51 (1.04)	407.77 (1.00)	422.45 (0.96)	436.19 (0.92)	450.50 (0.89)
4	Risk Margin	783.01 (2.08)	815.55 (1.99)	844.89 (1.92)	872.38 (1.84)	901.00 (1.78)
5	Managerial cost	783.01 (2.08)	815.55 (1.99)	844.89 (1.92)	872.38 (1.84)	901.00 (1.78)
6	Total Variable Cost	20129.87 (53.47)	21017.86 (51.41)	21686.70 (49.17)	22137.03 (46.79)	22532.50 (44.47)
7	**Fixed Cost**					
a)	Land revenue	31.00 (0.08)	31.00 (0.08)	31.00 (0.07)	31.00 (0.07)	31.00 (0.06)
b)	Depreciation	2399.15 (6.37)	2399.15 (5.87)	2399.15 (5.44)	2399.15 (5.07)	2399.15 (4.74)
c)	Interest on fixed capital	944.93 (2.51)	944.93 (2.31)	944.93 (2.14)	944.93 (2.00)	944.93 (1.86)
d)	Rental value of land	11775.00 (31.28)	11775.00 (28.80)	11775.00 (26.70)	11775.00 (24.89)	11775.00 (23.24)
e)	Interest on past capital	2363.77 (6.28)	4716.51 (11.54)	7271.78 (16.49)	10028.57 (21.20)	12985.80 (25.63)
	Total fixed cost	17513.85 (46.53)	19866.58 (48.59)	22421.86 (50.83)	25178.64 (53.21)	28135.87 (55.53)
8	Total cost	37643.71 (100.00)	40884.44 (100.00)	44108.56 (100.00)	47315.67 (100.00)	50668.37 (100.00)

Figure in parentheses are percentage to total cost

In case of overall farms, it was Rs. 38391.13 Rs. 40998.07, Rs. 44050.47, Rs. 47271.17 and Rs. 50540.90 per hundred plants in the first to fifth year, respectively (Table 5). Total variable cost in case of overall farms contributed 46.73 per cent in the first year and 57.67 per cent of total costs in the fifth year.

Table 5: Maintenance cost during non- bearing stage of mango orchard in case of overall farms

(Rupees/100 Plants)

Sr.No.	Particulars	Age of plantation in years				
		1st year	2nd year	3rd year	4th year	5th year
1	**Material Cost**					
a)	Gap filling	156.55 (0.41)	-	-	-	-
b)	Bordeaux mixture	189.68 (0.49)	234.57 (0.57)	255.70 (0.58)	293.31 (0.62)	437.74 (0.87)
c)	FYM	1483.09 (3.86)	1525.74 (3.72)	1607.60 (3.65)	1660.42 (3.51)	1645.47 (3.26)
d)	Plant protection chemical	324.12 (0.84)	362.49 (0.88)	374.50 (0.85)	387.52 (0.82)	396.16 (0.78)
	Total material cost	2153.43 (5.61)	2122.79 (5.18)	2237.81 (5.08)	2341.25 (4.95)	2479.37 (4.91)
2	**Human Labour**					
a)	Family labour	17255.96 (44.95)	17326.93 (42.26)	17595.35 (39.94)	17727.69 (37.50)	17698.39 (35.02)
b)	Hired labour	1752.55 (4.56)	1892.40 (4.62)	1954.66 (4.44)	2119.40 (4.48)	2256.94 (4.47)
	Total Human Labour	19008.50 (49.51)	19219.34 (46.88)	19550.01 (44.38)	19847.09 (41.99)	19955.33 (39.48)
3	Interest on working capital	195.30 (0.51)	200.76 (0.49)	209.62 (0.48)	223.03 (0.47)	236.82 (0.47)
4	Risk Margin	390.60 (1.02)	401.52 (0.98)	419.25 (0.95)	446.06 (0.94)	473.63 (0.94)
5	Managerial cost	390.60 (1.02)	401.52 (0.98)	419.25 (0.95)	446.06 (0.94)	473.63 (0.94)
6	Total Variable Cost	22138.43 (57.67)	22345.93 (54.50)	22835.94 (51.84)	23303.50 (49.30)	23618.78 (46.73)
7	**Fixed Cost**					
a)	Land revenue	31.00 (0.08)	31.00 (0.08)	31.00 (0.07)	31.00 (0.07)	31.00 (0.06)
b)	Depreciation	1662.01 (4.33)	1662.01 (4.05)	1662.01 (3.77)	1662.01 (3.52)	1662.01 (3.29)
c)	Interest on fixed capital	636.79 (1.66)	636.79 (1.55)	636.79 (1.45)	636.79 (1.35)	636.79 (1.26)
d)	Rental value of land	11775.00 (30.67)	11775.00 (28.72)	11775.00 (26.73)	11775.00 (24.91)	11775.00 (23.30)
e)	Interest on past capital	2147.91 (5.59)	4547.35 (11.09)	7109.73 (16.14)	9862.89 (20.86)	12817.33 (25.36)
	Total fixed cost	16252.70 (42.33)	18652.14 (45.50)	21214.52 (48.16)	23967.68 (50.70)	26922.13 (53.27)
8	Total cost	38391.13 (100.00)	40998.07 (100.00)	44050.47 (100.00)	47271.17 (100.00)	50540.90 (100.00)

Figure in parentheses are percentage to total cost

Total fixed cost varied from 42.33 per cent to 53.27 per cent of the total costs in the first year to fifth year .The material cost was increased with the age of plants. The interest on working capital showed the positive relation with age while interest on fixed capital decreased with age of plantation. The depreciation cost and rental value of land was declined with the increase age of plants. While interest on past establishment, cost was found to have positive relationship with the age during non-bearing years. Interest on past establishment, cost was increased with the increase in the age of plants which was the main component responsible for highest per cent share of fixed cost in the total cost.

Cost during bearing stage of mango plantation in different farm categories

Table 6 revealed that in marginal farm category, total cost increase in the early and main bearing stage and was calculated to Rs. 56373.26 and Rs. 58275.84 per hundred plants and then gradually decreased to Rs. 54354.65 in the late bearing stage. The fertilizers were maximum in main bearing stage Rs.2639.86 followed by late bearing stage Rs.2600.00 and early bearing stage Rs. 2419.35. The per cent share of FYM to total cost was maximum in main bearing stage (8.76%) followed by early bearing (8.56%) and late bearing stage (8.18%) respectively. Total labour cost increased from 36.46 per cent to 36.92, per cent from early to main bearing stage and thereafter, it decreased to34.34 per cent. The share of variable cost showed an increasing trend up to the main bearing stage and thereafter, it declined in the late bearing stage. However, fixed cost in percentage term to total cost exhibited declining trend up to the main bearing stage and then started increasing.

In small farmers, total cost varied from Rs. 55805.36 to Rs. 57532.91 per hundred plants from early to main bearing stage and then gradually deceased in Rs. 55832.37 per hundred plants respectively (Table 7). The per cent share of total material cost in total cost was maximum in main bearing stage (9.34%) followed by late bearing stage (9.26%) and early bearing stage (8.13%) in small farm category. Human labour varied from 39.80 per cent to 41.17 per cent from late bearing to early bearing stage. The share of variable cost in the

Table 6: Maintenance cost during bearing stage of mango orchard in case of marginal farms

(Rupees/100 Plants)

Sr.No.	Particulars	Bearing stages of mango		
		Early bearing	Main bearing	Late bearing
1	**Material Cost**			
a)	Bordeaux mixture	266.58 (0.47)	306.60 (0.53)	261.11 (0.48)
b)	FYM	4935.48 (8.58)	5000.00 (8.76)	4444.44 (8.18)
c)	Fertilizer	2419.35 (4.29)	2639.86 (4.53)	2600.00 (4.78)
d)	Plant protection chemical	641.94 (1.14)	667.36 (1.15)	635.56 (1.17)
	Total material cost	8263.35 (14.66)	8613.82 (14.78)	7941.11 (14.61)
2	**Human Labour**			
a)	Family labour	16572.94 (29.40)	15511.81 (26.62)	13222.22 (24.33)
b)	Hired labour	3982.97 (7.07)	6003.47 (10.30)	5444.44 (10.02)
	Total human labour	20555.91 (36.46)	21515.28 (36.92)	18666.67 (34.34)
3	Interest on working capital	612.32 (1.09)	730.86 (1.25)	669.28 (1.23)
4	Risk Margin	1224.63 (2.17)	1461.73 (2.51)	1338.56 (2.46)
5	Managerial cost	1224.63 (2.17)	1461.73 (2.51)	1338.56 (2.46)
6	Total Variable Cost	31880.85 (56.55)	33783.42 (57.97)	29887.24 (54.99)
7	**Fixed Cost**			
a)	Land revenue	31.00 (0.05)	31.00 (0.05)	31.00 (0.06)
b)	Depreciation	1180.31 (2.09)	1180.31 (2.03)	1180.31 (2.17)
c)	Interest on fixed capital	446.68 (0.79)	446.68 (0.77)	446.68 (0.82)
d)	Rental value of land	11775.00 (20.89)	11775.00 (20.21)	11775.00 (21.66)
e)	Pro-rated establishment cost	11059.42 (19.62)	11059.42 (18.98)	11059.42 (20.35)
	Total fixed cost	24492.42 (43.45)	24492.42 (42.03)	24467.42 (45.01)
8	Total cost	56373.26 (100.00)	58275.84 (100.00)	54354.65 (100.00)

Figure in parentheses are percentage to total cost

Table 7: Maintenance cost during bearing stage of mango orchard in case of small farms

(Rupees/100 Plants)

Sr.No.	Particulars	Bearing stages of mango		
		Early bearing	Main bearing	Late bearing
1	**Material Cost**			
a)	Bordeaux mixture	218.25 (0.39)	291.21 (0.51)	267.86 (0.48)
b)	FYM	1704.76 (3.05)	2032.97 (3.53)	2023.81 (3.62)
c)	Fertilizer	2092.38 (3.75)	2454.95 (4.27)	2345.24 (4.20)
d)	Plant protection chemical	523.81 (0.94)	591.87 (1.03)	535.71 (0.96)
	Total material cost	4539.21 (8.13)	5370.99 (9.34)	5172.62 (9.26)
2	**Human Labour**			
a)	Family labour cost	15422.22 (27.64)	13602.56 (23.64)	14444.44 (25.87)
b)	Hired labour cost	7555.56 (13.54)	9561.54 (16.62)	7777.78 (13.93)
	Total human labour	22977.78 (41.17)	23164.10 (40.26)	22222.22 (39.80)
3	Interest on working capital	604.74 (1.08)	746.63 (1.30)	647.52 (1.16)
4	Risk Margin	1209.48 (2.17)	1493.25 (2.60)	1295.04 (2.32)
5	Managerial cost	1209.48 (2.17)	1493.25 (2.60)	1295.04 (2.32)
6	Total Variable Cost	30540.67 (54.73)	32268.22 (56.09)	30567.69 (54.75)
7	**Fixed Cost**			
a)	Land revenue	31.00 (0.06)	31.00 (0.05)	31.00 (0.06)
b)	Depreciation	1719.48 (3.08)	1719.48 (2.99)	1719.48 (3.08)
c)	Interest on fixed capital	644.40 (1.15)	644.40 (1.12)	644.40 (1.15)
e)	Rental value of land	11775.00 (21.10)	11775.00 (20.47)	11775.00 (21.09)
f)	Pro-rated establishment cost	11094.81 (19.88)	11094.81 (19.28)	11094.81 (19.87)
	Total fixed cost	25264.69 (45.27)	25264.69 (43.91)	25264.69 (45.25)
8	Total cost	55805.36 (100.00)	57532.91 (100.00)	55832.37 (100.00)

Figure in parentheses are percentage to total cost

Total maintenance cost among different farm categories showed an increasing trend upto the main bearing stage and thereafter, it started decreasing. However, fixed cost in percentage term to total cost exhibited declining trend upto the main bearing stage and then started increasing. Maintenance costs during bearing stage of mango in medium farm category were presented in Table 8. Expenditure incurred on FYM and fertilizers increased from 3.91 per cent to 4.31 per cent and 4.18 per cent to 5.17 per cent in early and main bearing stage and gradually decreased in late bearing stage 4.25 per cent and 5.07 per cent, respectively. Total material costs varied from 9.89 per cent to 11.45 per cent of total cost of mango production. Interest on working capital increased in the early and main bearing stage and was calculated to Rs.756.76 and Rs. 917.94 and then decreased to Rs.Rs.843.50 in late bearing stage. The proportionate share of variable costs in the total maintenance cost showed increasing trends upto main bearing stage and thereafter, started decreasing. Among the fixed costs, rental value of land and prorated establishment cost accounted for maximum proportion of the fixed costs. The proportion of fixed costs exhibited almost declining trend with the age. Table 8 revealed that total cost varied from Rs. 55565.96 to Rs. 57547.18, respectively.

At overall level, total cost varied from Rs. 56873.29 to Rs. 57861.69 from early to main bearing stage (Table 9). The costs were decreasing in the late bearing stages in all the categories of farms. In the initial years of development stage of the plants, the labour and material inputs were increasing with the growth of the plants upto maturity in the main bearing stage and after that productivity of mango start decreasing. Expenditure incurred on fertilizers and plant protection chemicals ranged from 3.75 to 4.62 per cent and 1.01 to 1.17 per cent, respectively and the per cent share of total material cost in total cost ranged between 11.72 to 12.27 per cent exhibited positive relationships to age at overall level. Labour use almost doubled during the peak period of productivity phase i.e. in the main bearing stage as compared to early bearing stage and decreased in the late bearing stage. Table 4.24 revealed the per cent share of labour in total cost varied between 35.75 to 38.10 per cent in the various stages of mango. Risk margin and managerial costs varied from 2.35 per cent to 2.71 per cent at overall level and indicated positive relationship with age of plantation. Among fixed costs, rental value of land and prorated establishment cost accounted for maximum proportion of the fixed costs in all the categories of farms.

Table 8: Maintenance cost during bearing stage of mango orchard in case of medium farms

(Rupees/100 plants)

Sr.No.	Particulars	Bearing stages of mango		
		Early bearing	Main bearing	Late bearing
1	**Material Cost**			
a)	Bordeaux mixture	317.28 (0.57)	333.96 (0.58)	295.61 (0.55)
b)	FYM	2171.22 (3.91)	2481.20 (4.31)	2263.51 (4.25)
c)	Fertilizer	2324.48 (4.18)	2974.74 (5.17)	2702.70 (5.07)
d)	Insecticide	680.83 (1.23)	796.99 (1.38)	729.73 (1.37)
	Total material cost	5493.81 (9.89)	6586.89 (11.45)	5991.55 (11.24)
2	**Human Labour**			
a)	Family labour	10440.76 (18.79)	8417.54 (14.63)	6119.09 (11.48)
b)	Hired labour	9661.46 (17.39)	11771.93 (20.46)	10878.38 (20.41)
	Total human labour	20102.21 (36.18)	20189.47 (35.08)	16997.47 (31.89)
3	Interest on working capital	757.76 (1.36)	917.94 (1.60)	843.50 (1.58)
4	Risk Margin	1515.53 (2.73)	1835.88 (3.19)	1686.99 (3.16)
5	Managerial cost	1515.53 (2.73)	1835.88 (3.19)	1686.99 (3.16)
6	Total Variable Cost	29384.84 (52.88)	31366.07 (54.50)	27122.15 (50.88)
7	**Fixed Cost**			
a)	Land revenue	31.00 (0.06	31.00 (0.05	31.00 (0.06)
b)	Depreciation	2399.15 (4.32)	2399.15 (4.17)	2399.15 (4.50)
c)	Interest on fixed capital	944.93 (1.70)	944.93 (1.64)	944.93 (1.77)
d)	Rental value of land	11775.00 (21.19)	11775.00 (20.46)	11775.00 (22.09)
e)	Pro-rated establishment cost	11031.04 (19.85)	11031.04 (19.17)	11031.04 (20.69)
	Total fixed cost	26181.11 (47.12)	26181.11 (45.50)	26181.11 (49.12)
8	Total cost	55565.96 (100.00)	57547.18 (100.00)	53303.26 (100.00)

Figure in parentheses are percentage to total cost

Table 9: Maintenance cost during bearing stage of mango orchard in case of overall farms

(Rupees/100Plants)

Sr.No.	Particulars	Bearing stages of mango		
		Early bearing	Main bearing	Late bearing
1	**Material Cost**			
i)	Bordeaux mixture	271.20 (0.48)	309.16 (0.53)	280.35 (0.51)
ii)	FYM	3721.42 (6.50)	3448.71 (5.95)	3284.65 (5.88)
iii)	Fertilizer	2131.35 (3.75)	2672.29 (4.62)	2696.36 (4.86)
iv)	Plant protection chemical	574.02 (1.01)	678.74 (1.17)	650.19 (1.17)
	Total material cost	6697.99 (11.74)	7108.90 (12.27)	6911.55 (12.41)
2	**Human Labour**			
a)	Family labour	14995.79 (26.30)	13076.79 (22.58)	12137.51 (21.71)
b)	Hired labour	6670.10 (11.80)	8585.11 (14.86)	7740.99 (14.04)
	Total human labour	21665.90 (38.10)	21661.90 (37.44)	19878.50 (35.75)
3	Interest on working capital	668.40 (1.18)	784.70 (1.36)	732.63 (1.32)
4	Risk Margin	1336.81 (2.35)	1569.40 (2.71)	1465.25 (2.65)
5	Managerial cost	1336.81 (2.35)	1569.40 (2.71)	1465.25 (2.65)
6	Total Variable Cost	31705.91 (55.72)	32694.31 (56.50)	30379.92 (54.64)
7	**Fixed Cost**	0.00	0.00	0.00
a)	Land revenue	31.00 (0.05)	31.00 (0.05)	31.00 (0.06)
b)	Depreciation	1662.01 (2.94	1662.01 (2.88	1662.01 (3.02
c)	Interest on fixed capital	636.79 (1.13	636.79 (1.10	636.79 (1.16
d)	Rental value of land	11775.00 (20.71	11775.00 (20.35	11775.00 (21.22
e)	Pro-rated establishment cost	11062.59 (19.46	11062.59 (19.12	11062.59 (19.93
	Total fixed cost	25167.38 (44.28	25167.38 (43.50	25156.44 (45.36
8	Total cost	56873.29 (100.00	57861.69 (100.00	55536.36 (100.00

Figure in parentheses are percentage to total cost

Returns from Mango Cultivation

Table 10 depicts that the average production on marginal farm was 51.43 quintals during early bearing stage that increased to 55.70 during main bearing stage and then decreased to 51.88 quintal in late bearing stage and in case of small and medium farm, average production increased from 50.50 quintals to 54.07 quintals and 49.71 quintal to 53.33 quintals in early to main bearing stage and then gradually decreased to 50.79 quintals and 49.84 quintals in late bearing stage, respectively. It was observed that the average production was recorded highest in case of marginal farms i.e. 53.00 quintals followed by small farmers (51.79 quintals) and medium farmers (50.96 quintals). At overall level, the average production was 52.10 quintals.

Table 10: Farm category-wise average production from different age group of mango plants in the sampled households

(Quintals)

Farm category	Average production of mango /100 plants			
	Early bearing	Main bearing	Late bearing	Average production
Marginal	51.43	55.70	51.88	53.00
Small	50.50	54.07	50.79	51.79
Medium	49.71	53.33	49.84	50.96
Overall	50.70	54.59	51.02	52.10

The average gross returns were highest in case of marginal farms i.e. Rs. 185507.41 followed by small farms Rs. 181254.35 and medium farms Rs. 178365.54 (Table 10). At overall level, gross returns was Rs 177446.66 during early bearing stage that increased to Rs. 191064.70 in main bearing stage and then decreased to Rs. 178558.89 in late bearing stage. Average gross returns of mango at overall level were Rs. 182356.75

Table 11: Farm category-wise average gross returns from different age groups of mango in the sampled households

(Rupees)

Farm category	Average gross return of mango /100 plants			
	Early bearing	Main bearing	Late bearing	Average gross return
Marginal	180000.00	194957.98	181564.25	185507.41
Small	176750.00	189235.27	177777.78	181254.35
Medium	173987.27	186666.67	174442.68	178365.54
Overall	177446.66	191064.70	178558.89	182356.75

Summary

Initial cost of mango plantation per hundred plants was Rs. 34940.45 at overall level. Total variable and fixed cost was 59.13 per cent and 40.87 per cent, respectively of total cost. Maintenance cost during non-bearing stage per hundred plants at overall level ranged from Rs. 38391.13 to Rs. 50540.90 in the first year to fifth year, respectively. Interest on past establishment cost and rental value of owned land contributed more than 48.00 per cent of total cost. Maintenance cost of bearing mango per hundred plants at overall level showed an increase in cost upto main bearing stage and thereafter, it started decreasing which is due to the decrease in production and lead to lesser labour requirement. The proportion of variable cost ranged from 55.72 per cent to 56.50 per cent upto main bearing stage. The per cent share of fixed cost varied from 43.50 per cent to 45.36 per cent in the total cost in different ages of plantation. Cost incurred on material inputs ranged from 11.74 per cent to 12.41 per cent while human labour cost was around 35.75 per cent to 38.10 per cent of the total cost for the different age groups. Average production varied from 50.96 quintals to 53.00 quintals per 100 plants and net returns varied from Rs. 178365.54 to Rs. 185507.41 in different farm categories.

References

1. National Horticulture Board, (2016). *Horticultural Data Base, 2016.* National Horticulture Board, New Delhi.
2. Directorate of Economics and Statistics, (2016). *Statistical Outlines of Himachal Pradesh, 2016*. Directorate of Economics and Statistics, Government of Himachal Pradesh.
3. Chand R. (1994). Economics of perennial crops: Some methodological issues, *Indian Journal of Agricultural Economics* 49: 1420–30.
4. Akcay Y. and Uzunoz M. (2005). An investment analysis of peach and cherry growing in Turkey, *Journal of Applied Science* 5: 1665–68.

❑❑❑

Chapter 8

Research Methodology for Social Sciences

Manpreet Kaur and Dr. Anil Bhat

Division of Agricultural Economics and ABM, SKUAST-Jammu

Introduction

Human beings are the most intelligent social animals who are known to modify the external environment and conditions according to them. After fulfilling the basic needs of life, everyone is working to live life with ease and for that a number of researches have been ongoing in every field. But as long as the research findings are not beneficial to masses or they are out of masses league, the researches are not utilized as much they should. So, to understand whether those researches are known to effect our masses, social science research is important and if it is not affecting the masses, the reason as to why it is not being accepted by the masses is also studied in social science research where different patterns of behaviour, human culture, customs are studied and inferences are drawn. There has been a definite pattern of behaviour which have been expected from the humans whether it comes to technology adoption, perception of new technology, etc. which has been framed according to ages of continuous research and a lot more is needed. This chapter includes an introduction to research process, paradigms of the research process, overview of research process, reporting of the research, theories and model framing in research and problems in social research.

According to P.V. Young: "Social research is a scientific undertaking which by means of logical methods, aim to discover new facts or old facts and to analyze their sequences, inter-relationships, casual explanations and natural laws which govern them."

According to C.A. Moser, "Social research is a systematize investigation to gain new knowledge about social phenomena and problems."

Characteristics of Social Research

- Its ultimate goal is to find cause and effect relationship.
- It emphasize on development of generalization (principles or theories) by which upcoming events could be predicted.
- It is based on observable facts and realistic evidences.
- It involves collecting data from primary sources and using secondary information from existing sources.
- It needs careful recording and reporting. Every important term is defined, restraining causes are recognized, procedures are described, references are acknowledged, results are detailed and conclusions are presented.
- It is inter-disciplinary in nature.

Objectives of Social Research

- To comprehend the human behaviour.
- To standardize the concepts of society.
- To frame explanations to social problems.
- To acquire knowledge about social phenomena.
- To find natural laws that governs social phenomena.

Types of Research

From application point of view

1. Pure research: It involves testing theories and hypothesis which appear challenging to the researcher and may not have practical application in the real world.
2. Applied research: The techniques or methods of research when applied to the real world for collecting data and information which can be further used for policy framing and administration is the applied research.

From objective point of view

1. Descriptive study: It basically deals with describing the research problem or our idea of interest. The main idea behind this type of research is that a problem well-defined is a problem half solved.

2. Correlational study: This type of study deals with establishing a relationship between two or more aspects of a situation and tries to explain how one aspect of a particular situation is affecting the other aspects.
3. Exploratory study: It is also known as pilot study or feasibility study, as it is a very limited study performed on a very small scale to check whether a particular research study can be undertaken or performed in that area and according to this study, a full fledged study may eventuate.
4. Explanatory study: This type of study deals with clarifying the relationship between different aspects of a phenomena.

Depending on the mode of enquiry

1. Quantitative study: It is experimental of experiential examination or study of phenomena which can be observed by using statistical, mathematical or computational techniques. In this research, data is collected from different resources and then inferences are drawn by analyzing it carefully.
2. Qualitative study: It is exploratory study which provides vision or understanding into problems of the research area and helps to develop hypothesis or theories for testing in quantitative research.

The cycle of Research

The theories and observations are two pillars of scientific research which operates at two level: theoretical level and empirical level. Theoretical level is concerned with building hypothesis or theories which are tested at the empirical level. Therefore, a scientific enquiry possibly can take two forms: inductive research where the researcher is making theoretical concepts according to the observed observation and deductive research where the researcher test theoretical concepts using new empirical data. Hence, inductive research is called as theory building research and deductive research is called theory testing research.

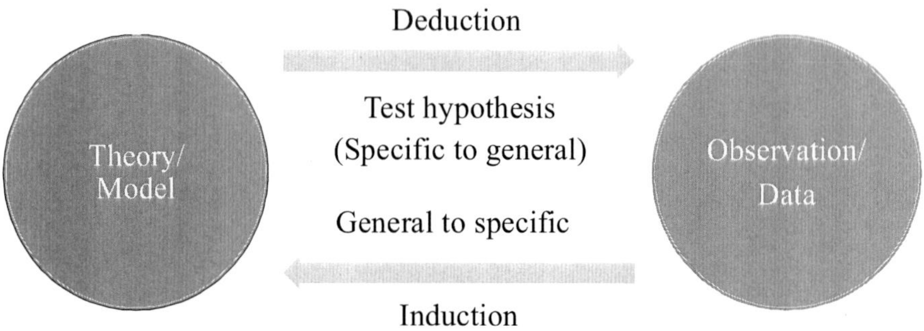

However, in social sciences, both theory building and theory testing are difficult tasks due to ambiguous nature of theoretical concepts, inadequate tools to measure them and presence of many other unobservable factors that also affect the phenomena of interest. The theory once credited as being effective, couldn't be suited to all the conditions and may termed inferior in other situations or may lose its credibility with time. Therefore, for scientific research, two different skills are required: theoretical and methodological. Methodological skills can be easily acquired and relatively standard while theoretical skills can not be taught and are learned through experience.

Paradigms of Social Research

The mental models or belief system by which one organize one's reasoning and observations are recognized are called paradigms. The word "paradigm" was first used by Thomas Kuhn (1962) in his book "The Structure of Scientific Revolutions" in which he examined the history of natural sciences to identify patterns of activities that shape the progress of science. As the social reality is viewed differently by different people, which constrain their thinking and reasoning about the observed phenomena.

The popular paradigms in social science research are:

1. Positivism : It was based on the works of French philosopher Auguste Comte and was dominant until mid- 20th century. According to this, scientific research should only be limited to what we can observe and measure which means it relies completely on theories that can be tested. Positivism was originally an attempt to separate scientific research from the religion but later it led to empiricism, which means blind faith in observed data and any attempt to reason beyond the observable

facts are rejected. As human behaviour, thoughts and emotions can not be measured, they were not considered to be reasonable topics for scientific research and it led to the development of post-positivism.

2. Post-Positivism/Post-Modernism : According to post-positivism, the reasonable inferences about a particular phenomenon are drawn by combining empirical observations with logical reasoning. Post-positivist view science as probabilistic based on many possibilities and uses them to understand social phenomena better. Post-positivists are further divided into subjectivists who view the world as construction of our subjective minds rather than objective reality, and critical realists, who believe of an external reality independent of a person's thinking, which can never be known with certainty.

Overview of the research process in social sciences

All scientific research is an iterative process of observation, rationalization and validation:

- In observation phase, a social phenomenon, event or behaviour of the interest is observed by the researcher.
- In rationalization phase, a theory or hypothesis is made according to the observations by logically linking the different bits of the problem.
- In validation phase, hypothesis or theories constructed in the rationalization phase are tested using a scientific method and in this process the initial theory is possibly modified or extended.

Most traditional research tends to be deductive and functionalistic. The figure below provides a series of activities to be performed in functionalistic research:

1. **Exploration :** It includes exploring and selecting research problems for further study, examination of the published literature of related studies done and identification of theories which may be able to explain the research problems of interest. It includes:

a) Identifying research problems : The research problems are of two types: one which relates to the states of nature and other which relates to relationship between variables. The problem first may be stated in a broad general way but the ambiguities are later removed and the feasibility of a particular solution has to be considered prior to working on it. The formulation of a specific research problem from a general topic is the identification of research problem.

b) Literature review : Its purpose is to know the current state of information about the study, to identify key authors, articles, studies and findings in that area and to identify gaps in knowledge in the study area. The literature review would indicate that whether research problems of our interest have already been addressed in the literature or they should be modified according to the literature review. The literature also provides possible answers to our research problem and helps us to recognize theories which have already been used to address similar research problems.

c) Theory testing : The theories which can help to address the research problem are identified in this step. A theory basically helps to identify the concept logically answering the research problem. In functionalistic research, theories are used for hypothesis formulation for empirical testing.

2. **Research design** : This is concerned with making a plan of the activities which is to be followed later in order to reasonably answer the research problem identified in the exploration phase. It includes:

a) Operationalization : The theoretical theories which are made in the previous step of exploration are measured in operationalization, which refers to developing indicators for measuring those theories. As there is high level of bias, inaccuracy and prejudice in social science research, a number of indicators are used to measure most of the theories. For example, if we have to measure socio economic status of a person, it can be operationalized using a number of indicators, such as what is your family income, what are your farm assets, what are your livestock inventory, etc. The indicators when combined at empirical level to represent a theory are known as variable which can be dependent, independent, mediating or moderating based on how it is used in the study. An indicator may have several attributes representing values.

And value of these attributes may be qualitative and quantitative in nature. In social science research, even if the variables are qualitative, they are tried to be presented as quantitative. For examples, to indicate whether the farmers think the government policies are helpful to them, an indicator with three attributes can be made; agree, neutral, disagree.

b) Research method: Along with operationalization, the researcher also decides the method he would employ to collect data for answering his research problem. Research method can be quantitative, such as experiment, survey research or qualitative such as case study or a combination of both.The experimental design is designed in this step, if it needs to be conducted. If the survey is to be done, it is decided that whether it would be a questionnaire which is to be mailed later or a schedule which is to be interviewed personally later by the researcher.

c) Sampling strategy: After selecting the research method, target population from which the data is to be collected is carefully chosen. Social science research basically deals with inferring behavioural patterns within specific population and the entire population can not be studied or surveyed due to time limit, lack of money, therefore a sample representing the whole population is studied. The sampling is done in three steps: defining the target population, selection of sampling frame, drawing of sample.

At this stage, a research proposal or synopsis having all the decisions made in the previous stages are written and finalized. This multi-part proposal consist the research problem, which is to be answered, the prior information related to the research problem, theories which are to be tested, indicators to measure the theoretical concepts, research method to be employed and sampling strategy. This research proposal acts as means to get feedback from others and identifying the lacunas in the plan before the research execution which can be incorporated easily in this stage. This type of research proposal is also obligatory by funding agencies to finance the research.

3. **Research execution**: This step is the final step of the research and it includes:

a) Pilot testing : This is also known as pre-testing of the schedule in which a small subset of sample population, which is also known as pilot sample is tested with the prepared questionnaire or schedule and it is viewed whether the collected information by this method is enough to answer the research problem. The unwanted questions can be removed and required can be added in this step.

b) Data collection : The data is collected from the target population (sample) with the help of pre tested schedule or questionnaire. It can be done by personally interviewing the person or by mail, telephone, etc.

c) Data analysis : In the final step, the data is analyzed which requires establishment of categories, application of these categories to raw data through coding, tabulation and then drawing statistical inferences. The researcher needs to condense raw data into manageable groups by coding, editing and tabulation. In coding, the categories of data are altered into symbols which can be tabulated and counted later. Editing is the removal of unnecessary data to improve the quality of data and tabulation is placing of classified data into tabular form. After tabulation, data is analyzed by applying well defined statistical formulae.

The last step of the research is the research report documentation in which all the research findings are documented. The writing of research report must be done keeping in view the layout of report which contains preliminary pages, main text and the end matter. The title, acknowledgements, foreword are written in the preliminary pages which are followed by a list of tables, list of graphs and charts. The main text of the report is divided into following parts:

1. Introduction : It contains the objectives of the research along scope and importance of study along with limitations is also stated in this.

2. Review of literature : The findings of other researchers who have performed similar researches are documented in this part.

3. Materials and methods : The methodology adopted during the study was written in this part.

4. Results : The results obtained in the research are documented in this part in technical manner in detail.

5. Discussion : The results obtained in the previous part are discussed in this part and the patterns of results obtained are studied and cause and effect relationship are established by giving explanations to the research obtained.
6. Summary and conclusion : The results obtained are summarized and concluded in this part.

At the end of the report, bibliography or references of books, journals, research papers, journals, reports which were consulted during the study are documented followed by appendices.

Theories and Models in Social Research

A theory is a collection of related concepts or prepositions to explain a phenomenon within certain assumptions and boundaries. Theory must be well-supported by the observed facts to validate it. As Kurt Lewin once said," Theory without practice is sterile; practice without theory is blind". Hence, theory supported by facts is essential for scientific research. Nevertheless, the theory is not the truth and must not be accepted as it was proposed by someone and is accepted and valid at a point of time because with course of time, poorer theories are eventually replaced by better theory which were more explanatory. So, the basic requirement of the research is to build better and more comprehensive theories explaining a phenomenon better than the previous theories.

A model, on the other hand, is representation of the phenomena. A theory explains a phenomenon while a model tries to represent it. Models may be of different kinds, such as mathematical models, network models and path models. The process of model or theory development could be deductive or inductive which go hand in hand in it. The following model building process has been given by Amol Bhattacherjee (2012).

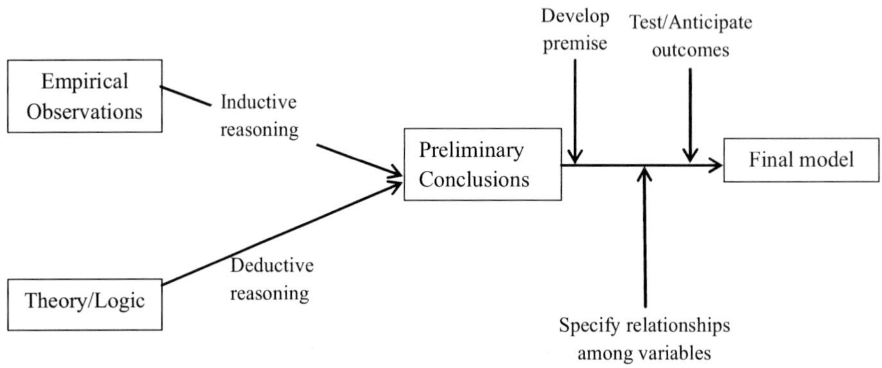

The model building process

Problems in Social Science Research

1. The complex nature of social data due to dependence of human behaviour on several factors, such as physical, social, temperamental, psychological, geographical, biological, socio cultural, etc. It is because of this reason, human beings can not be put to scientific tests.

2. Social scientists have to face the problem of abstraction and faulty reasoning due to which wrong conclusions can be drawn. For example, According to classical theory of employment, wage cut will increase employment opportunities however, later J. M. Keynes provides the general theory of employment.

3. Problem in inferring cause and effect relationship between variables are found in some cases. For example, In under-developed economies, the human development index can not be improved due to lack of basic resources and capital to provide basic resources can not be provided due to the under-developed economy.

4. Dynamic nature of social phenomena as human societies undergoes constant change. The techniques used in the past may prove useless today and due to this, the task of analyzing data becomes complicated and wrong inferences can be drawn.

5. Problem of preserving fairness due to subjective nature of social research. As the social scientists have their own bias and prejudices, which are reflected in their findings because of the less objective nature of the social phenomena.

6. The social scientists can not easily predict the social data with high degree of accuracy, which can only be improved with experience.
7. There is difficulty in verification of inferences as social events are non-repetitive in nature.
8. There is a difficulty in use of experimental method in social science research because human beings can not be put to laboratory testing and even if it is done, their responses wouldn't be natural.

References

1. Bhattacherjee, A. (2012). *Social Science Research: Principles, Methods and Practices,* pp. 4-24. Textbooks collections, Florida.
2. Kothari, C.R. (2004). Research Methodology: Methods and Techniques, pp. 7-24. New Age International Publishers, New Delhi.
3. Kumar, R. (2011). *Research Methodology: a step by step guide for beginners.* Sage Publications, New Delhi.

❑❑❑

Chapter 9

Investigating the Role of Mediation Analysis in Social Sciences Research

Dr. Shelleka Gupta

Assistant Professor, The Business School, University of Jammu, (J&K)

The most interesting and prominent research questions in social sciences are about statistical relationships between various variables. Based on the review of literature and data analysis, variables may be grouped as dependent or independent. Data analysis usually focuses on examining if the presumed causal variable, whether manipulated or measured, is related to the outcome using a linear model, such as analysis of variance or linear regression. In many arenas of research, such analyses, when accompanied by good research design, are sufficient to answer the question as to whether variation in *X* causes variation in *Y*. But deeper understanding accrues, when researchers investigate the process by which a given effect is produced.

There is a need to understand more about how these effects or relationships arise. The mechanism that produces and explains the associations are mediation—the process through which *X* exerts its effect on *Y* through one or more mediator variables. Mediating variables are common in social sciences research, as they help in understanding the relationships among variables more clearly. In the simplest mediation theory, the investigation of mediation specifies a chain of relations by which an antecedent variable affects a mediating variable, which in turn affects a dependent variable. Mediating variables is one that transmit the effect of one variable to another variable.

If a variable *M* is causally situated between *X* and *Y* and accounts for their association (at least in part), it is said that *M* mediates the relationship. It is also said that *X* has an indirect effect on *Y* through *M*. Mediation analysis is the process of determining whether or not variables acting as an in-between step, called mediators, are present when looking at the relationship between an independent variable X and a dependent variable Y.

Simple Mediation

In simple mediation model, there is only a single proposed mediating variable. The fig. 1 represents a causal process in which *X*, the independent variable, affects *Y*, the dependent variable. Path *c* quantifies this effect called the total effect of *X* on *Y*. Path *a* represents the causal effect of the independent variable on the proposed mediator, *M*. Path *b* represents the causal effect of the mediator on the dependent variable, controlling for the independent variable, whereas path *c*2 represents the causal effect of the independent variable on the dependent variable controlling for the mediator. In the language of causal analysis, *c*2 is the direct effect of *X* on *Y* and is distinguishable from the total effect, *c* in that the direct effect partials out from the total effect that part of the causal effect that is shared with *M*. That is, it represents the part of the effect of *X* on *Y* in the model that is unique to *X*. Path *b* can also be considered a direct effect, in this case the direct effect of the mediator on the outcome variable. The indirect effect of *X* on *Y* is represented as the two paths linking *X* to *Y* through *M*, which in Figure 1 are the a and *b* paths.

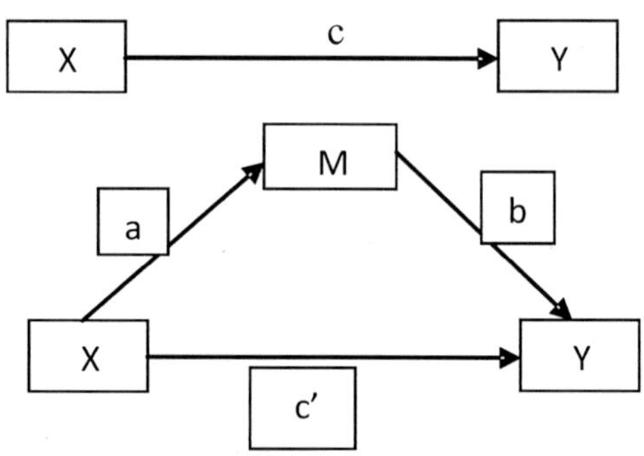

Figure: 1 A simple mediation model
Source: Preacher and Hayes (2007)

Multiple Mediation

Variables often exert their effects through multiple mediators. Multiple mediation improves our ability as researchers to identify, model, and predict outcomes with complex antecedent Chains. A graphical depiction of a model involving multiple mediators can be found in Figure 2.The top panel of the figure represents the total effect of X on Y, represented with the Unstandardized path coefficient c, and the bottom panel represents both the direct effect of X on Y (path c2) and the direct effects of M on Y (the b paths). The specific indirect effect of X on Y via mediator j is defined as the product of the two unstandardized paths linking X to Y via that mediator. For example, the specific indirect effect of X on Y through M_1 is quantified as a_1b_1. The total indirect effect is the sum of the specific indirect effects, $\sum i(aibi), i=1$ to j, where j is the number of proposed mediators. The total effect of X on Y is the sum of its direct effect and the j specific indirect effects; that is, $c = c' + \sum i(aibi), i=1$ to j.

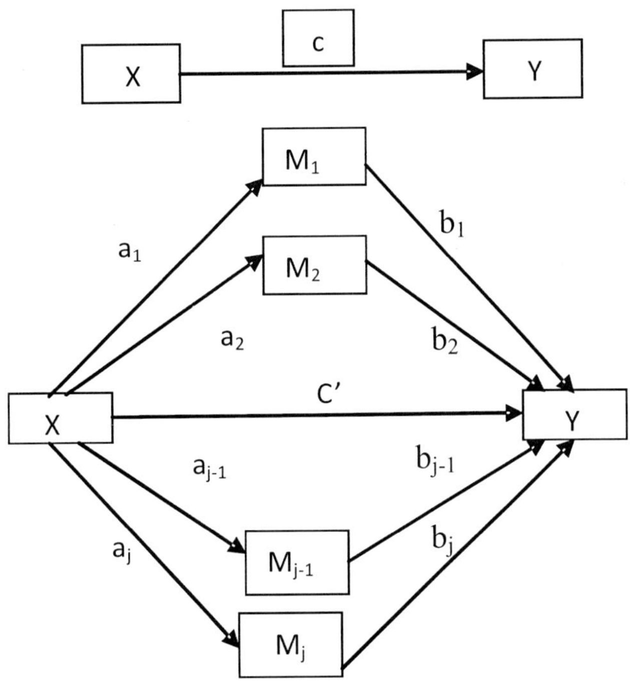

Figure 2: A Single-Step Multiple Mediator Model
Source: Preacher and Hayes (2007)

The two-mediator model helps to compare the causal steps, difference in coefficients, and product of co-efficients approaches to testing mediation in more complicated models. When the mediation model includes two mediators, the regression equations can be expanded as follows (MacKinnon, 2008):

$Y = i1 + c X + e1,$

$Y = i2 + c' X + b1M1 + b2M2 + e2,$

$M1 = i3 + a1X + e3,$

$M2 = i4 + a2X + e4,$

Where X is the single independent variable, M1 and M2 are two mediators, and Y is the dependent variable. The co-efficients a1 and $a2$ represent the relations between the independent variable and the two mediators, respectively. The co-efficients b1 and b2 represent the relations between each mediator and the dependent variable controlling for the independent variable and the other mediator. Again, an estimator of the mediated effect can be obtained by the difference between the total effect and the direct effect, that is, c- c′ or by the product of the relevant co-efficients (i.e., a1b1 and *a2b2)*. The estimator of c- c′ is the total mediated effect, that is, the mediated effect via both M1 and M2. On the other hand, the estimators of a1b1 and *a2b2* are the mediated effects uniquely attributable to M1 and M2, respectively. In the product of co-efficients method, however, it is much more straightforward to test whether the total mediated effect, the sum of a1b1and *a2b2,* is statistically significant either by forming the ratio of the estimate to estimated standard error or by creating confidence limits via re-sampling method. Effect size in mediation models can be specified for each path involved in the mediation pathway and for the entire mediated effect. To specify the effect size of each path, correlations, partial correlations, and standardized co-efficients are used.

Complete Versus Partial Mediation

The total effect *c* is the sum of the mediated effect *(c* - c′ or *ab*) and the direct effect *c′*. In case of complete mediation, the total effect is completely explained by the mediator and also, when the mediated effect or indirect effect is statistically significant and the direct effect is not statistically significant. But in case of partial mediation, the relation between the independent and the dependent variables is not completely explained for by the mediator. In this case both the direct effect and the mediated effect are statistically significant, reflecting that the mediator significantly explains the relationship between the independent and the dependent variables.

Illustration : To check service quality (SQ) and perceived risk (PR) as variables that mediates the relationship between Cause Related Marketing (CRM) and Customer Value (CV). It is multiple mediation model and the mediation analysis is done by using PROCESS that is a regression-based path analysis macro available for both SPSS and SAS that estimates the model co-efficients in mediation and moderation models of various forms while also providing modern inferential methods for inference about indirect effects including bootstrap confidence intervals.

Table 1: Total Effect and Direct Effect of Independent Variable on Dependent Variable.

	Total Effect of Independent Variable on Dependent Variable			
	coeff	**Se**	**T**	**p**
CRM	**0.6421**	**0.0335**	**19.1665**	**0.0000**
	Direct Effect of Independent Variable on Dependent Variable			
	coeff	**Se**	**T**	**p**
CRM	.4428	.0344	12.8632	.0000

Source: Authors calculations

The total effect of independent variable on dependent variable taking into consideration mediators is significant as p=0.000. And the direct effect of independent variable on dependent variable (without taking into consideration Mediators) is also significant as p=0.000. It can also be observed from the above table that the inclusion of mediators in the relationship of independent and dependent variable increase the effect from 0.4428 to 0.6421.

Table 2: Indirect Effects of Independent Variable on Dependent Variable through Proposed Mediators

	Effect	Boot SE	Boot LLCI	Boot ULCI
TOTAL	.1994	.0246	.1556	.2493
SQ	.1983	.0247	.1535	.2491
RISK	.0011	.0022	-.0011	.0100
(C1)	0.1971	0.0250	0.1517	0.2483

Source: Authors calculations.

In this case the true total indirect effect is estimated to be **0.1994** at 95% confidence and it lies between 0.1556 to 0.2493. The estimated total indirect effect is the difference between total effect and the direct effect (i.e. 0.6421–0.4428 = **0.1993**). The total indirect effect is significant as zero does not occur between the lower limit and the upper limit. Thus, it can be concluded that Independent Variable on Dependent Variable effects directly as well as indirectly via mediators.

Furthermore, it is clear from the above table that the total indirect effect via service quality at 95% confidence is 0.1983 and lies between .1535 and .2491. As in case of service quality, the zero does not lies between the lower limit and the upper limit, therefore it is concluded that the indirect effect for this mediator is significant. But, the indirect effect of perceived risk is estimated to be 0.0011 at 95% confidence and it lies between - 0.0011 to 0.0100. The negative sign in lower limit and a positive in upper limit signal the occurrence of zero between the lower limit and the upper limit, thus indicating that the role of perceived risk as a mediator, in the relationship between Cause Related Marketing and Customer Value is not significant.

Furthermore, the above table also represents the comparison of indirect effects of service quality and perceived risk. As the comparison (C1) does not contain zero, thus it leads to the conclusion that there is significant difference between the indirect effects of service quality and perceived risk. The indirect effect via service quality is greater than the effect via perceived risk. The above analysis highlights that Cause Related Marketing indirectly enhances customer value through service quality as compared to perceived risk.

Table 3: Model summary

Model Summary without mediators					
R	**R-sq**	**F**	**df1**	**df2**	**P**
.5249	.2755	367.3565	1.0000	966.0000	.0000
Model Summary with mediators					
R	**R-sq**	**F**	**df1**	**df2**	**P**
.6212	.3859	201.9047	3.0000	964.0000	.0000

Source: Authors calculations.

The impact of mediating variables in the relationship of Cause Related Marketing and Customer Value is visible from the above tables where the R square value rises from 0.2755 to 0.3859.

The above illustration helps the researcher in analysing the relationships between variables with great depth. The direct, indirect and total effect, depict the relevance of the proposed mediators that facilitates the understanding of problems in social science research. manifestation of mediation relations is a complex process, and thus, be based on thorough literature review. Apart from regression based analysis, there are other methods, such as Structural Equation Modelling (SEM), experimentation research have been in use for mediation analysis. There is need to apply mediation analysis in social sciences research to further strengthen its reliability and specificity of theoretical mediating concepts.

References

1. Preacher and Hayes, (2007). Contemporary approaches to assessing Mediation in Communication Research. Retrieved from: http://www.sagepub.in/upm-data/23657_Chapter2.pdf.

2. MacKinnon, D.P. (2008). *Multivariate applications series. Introduction to statistical mediation analysis,* New York, NY,: Taylor & Francis Group/Lawrence Erlbaum Associates.

❑❑❑